SIX QUANTUM PIECES

A First Course in Quantum Physics

SIX QUANTUM PIECES

A FIRST COURSE IN QUANTUM PHYSICS

Valerio Scarani
Centre for Quantum Technologies and Department of Physics
National University of Singapore, Singapore

Chua Lynn & Liu Shi Yang
NUS High School of Mathematics and Science, Singapore

Cover and illustrations by Haw Jing Yan

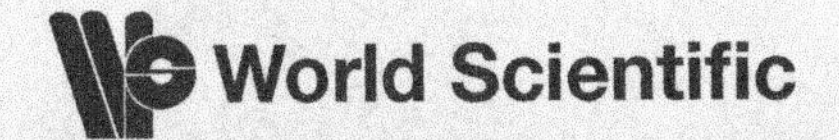

NEW JERSEY • LONDON • SINGAPORE • BEIJING • SHANGHAI • HONG KONG • TAIPEI • CHENNAI

Published by

World Scientific Publishing Co. Pte. Ltd.
5 Toh Tuck Link, Singapore 596224
USA office: 27 Warren Street, Suite 401-402, Hackensack, NJ 07601
UK office: 57 Shelton Street, Covent Garden, London WC2H 9HE

British Library Cataloguing-in-Publication Data
A catalogue record for this book is available from the British Library.

First published 2010
Reprinted 2011

SIX QUANTUM PIECES
A First Course in Quantum Physics

ISBN-13 978-981-4327-53-4
ISBN-10 981-4327-53-0
ISBN-13 978-981-4327-54-1 (pbk)
ISBN-10 981-4327-54-9 (pbk)

Desk Editor: Ryan Bong

Printed in Singapore.

Contents

Foreword

Many notions of modern science have found their way into school programs all over the world. The generations of students going through high school these days are receiving a good overview, and at times much more than an overview, of such exciting topics as the structure of the atom, the incredible complexity of the cell, the expansion of the universe ...

But this is not the case for quantum physics, an unrivaled body of knowledge, both in the precision of its predictions and in the breadth of its scope. Since the early decades of the twentieth century, quantum physics has been shaping modern science, from its deepest foundations to its most technological applications. Moreover, popular science books and magazines mention quantum physics regularly, because most of the research in physics is related to it. In spite of all these, school programs cautiously keep away from quantum physics. Why is this so?

There is an obstacle in the presentation of quantum physics, and a huge one at that: *one cannot make an image of quantum phenomena.* An atom can be drawn and the sketch, though inaccurate, conveys something; the DNA double helix can be drawn and it conveys even more information about the reality; stars can be seen with the naked eye, images of galaxies can be downloaded. It goes even further. Even if the earth looks flat to us walking on it, one could mentally have an image of what a round earth means even before pictures were taken from space. Even if the sun seems to be moving around us, one can imagine the earth rotating around the sun and even create nice animations of the solar system.

Take now some typical quantum facts: a unique particle can be delocalized in different regions of space; the position and momentum of a material object are never sharply defined; even more unthinkable, the fact called "entanglement": many objects can lose all their individual properties and share only a common set of properties. These are the core facts of quantum physics. There is no image for any of these. Let us skip here the deep implications of this fact and concentrate rather on the pedagogical problem: how can one learn quantum physics?

The last decades of research have opened a possible path around the obstacle. Several experiments have shown the unmistakable signature of quantum facts. The facts are still impossible to draw, but the experimental schemes can be drawn. Moreover, their results are just lists of numbers that can be read and analyzed.

This very direct contact with quantum facts has already had a momentous consequence. Some decades ago, it was still possible to believe with Einstein that quantum *theory* was weird, but that this was just due to the use of the wrong mathematical tools, to the physicists' lack of fantasy. It was still possible to hope that, sooner or later, a more "reasonable" presentation would be found. Nowadays, we know beyond any possible doubt that this dream is hopeless: it is *nature itself* that is weird, the theory is just adapted to the inescapable reality.

So, if you want to learn quantum physics, there is no reason to delay any longer. For those who choose to make the effort (because effort it will take), we have tried to make the path as straight as possible: a direct jump into the "real stuff" supported by some not too difficult mathematics. This text will not make you an expert of quantum physics: it is just meant for you to have a glimpse of the unexpected and fascinating possibilities that nature has in store.

We would like to conclude this foreword by each addressing our own colleagues.

From Lynn and Shi Yang to high school students and other beginners

Quantum physics is indeed one of the most difficult topics to learn and to understand. In this book, we have tried to present the material such that it can be easily understood. However for those who may still have problems, we offer the following advice.

First, the main trouble you may face is the notations and formalisms used in this book. The Dirac notation for vectors, the tensor product etc. may seem completely unfamiliar to you. However, do not be put off by these: it only takes a while to get used to these notations, and as you read further they may start becoming natural to you.

Second, at the high school level, you may still not be familiar with some of the concepts presented here, such as the vector space, transformations, information theory ... For these, we have included more information at the end of the respective chapters, under the section "The Broader View". A note of caution: some of the information presented under these sections may be difficult for you to grasp. If so, just focus on the main text; these sections are actually optional, extra material for the interested reader, and are not crucial to the main content of this book. The same goes for the last two chapters, under the part "Beyond the Six Pieces", which are purely for interested readers who would like to explore further.

Third, do take your time to work through most, if not all, of the exercises in this book. While we have written the book such that the text can stand alone without the exercises, working through them can actually help you to consolidate your understanding. We have also kept their difficulty at an acceptable level: most of the exercises should be straightforward and do not require difficult mathematics or problem-solving skills.

Finally, we hope that you would enjoy reading this book, and that this book would succeed in its purpose: to allow beginners to understand the basic concepts of quantum physics. Hopefully it would also compel you to further your explorations in the quantum world!

From Valerio to teachers

The challenge of presenting quantum physics in the high school is exactly this: a challenge! Exciting, but open to failure. At the moment of writing, I am not advocating that quantum physics becomes part of the general syllabus: this may prove to be too much. But for special classes or some particularly motivated students, I am convinced it is worth a try.

The pedagogical challenge is the following: *to present the students with something that cannot be fully grasped*, i.e., for which no adequate image can be drawn. Allow me some remarks on this crucial point.

First, there is a misunderstanding to be dispelled. The difficulty of grasping quantum physics is sometimes attributed to "the lack of mathematical tools". The argument misses the point completely. It misses it as the generic claim that one needs to master the mathematics before speaking about the world — indeed, as a counterexample, think of the illustration of curved space-time obtained by placing a heavy ball on a plastic sheet and letting a much lighter ball move around it: it is a perfect image of what it is supposed to mean, students can grasp its meaning without any knowledge of differential geometry. It misses the point also as a specific claim — just ask some university students if they have the feeling of having understood quantum physics, after having learned about operators, Hilbert spaces and path integrals. The difficulty in understanding quantum physics is deeper; at the same time, it is almost unrelated with the formalism: the mathematical tools can fortunately be kept simple.

Second, one should avoid conveying the message as "nobody understands quantum physics". This would be very derogatory for physicists and for physics as a discipline. Actually, some of us understand quantum physics so well, that we end up accepting its serious consequences: namely, the fact that some questions should not be asked, because nature definitely does not have the answer. This is the precise meaning of the statement "nature is weird" used earlier; this is the main message to be conveyed. Now, such a level of understanding can be reached only by personal inquiry. When we teach quantum physics, we must leave the students free for trying and finding all possible ways out of the right explanation; then, we have to reply by showing the experiment that falsifies their view, not by invoking Heisenberg or Bohr or Feynman.

A final remark concerning this specific text: I am almost certain that the material covered here has little to do with the way you learned quantum physics during your university days. You may even find the whole approach disconcerting, because it does not fit the format that you were expecting. With this in mind, I have written a special last chapter in order to bridge the gap between the usual presentations and this one.

Last but not least, do not hesitate in contacting me with questions, criticisms, suggestions: the challenge is still ongoing, this text is probably a first step rather than the ultimate summit.

Acknowledgments

We are especially grateful to Mr. Tan Kian Chuan, the head of department of Physics in NUS High School (Singapore), for having encouraged and promoted this project from the start.

Mr. Sze Guan Kheng, the head of department of Physics in Raffles Institution (Singapore), has given us very detailed feedback on an early version of the manuscript.

During the writing of this text, Lynn and Shi Yang have benefited from clarifying discussions with Lana Sheridan, Le Phuc Thinh, Charles Lim, Melvyn Ho and Wang Yimin (Centre for Quantum Technologies, NUS, Singapore).

This approach to quantum physics for high-schoolers underwent its first test in Collège St-Michel, Fribourg (Switzerland), back in 2005. Valerio thanks Mr. Patrick Monney and his students Ricardo Barrios, Benjamin Fragnière, Etienne Kneuss, Gaël Monney, Nguyen Thanh Hieu and Quentin Python.

A User's Guide

Main material

The main material is contained in chapters 1-7, excepting the sections "The Broader View" and "References and Further Reading", which can be skipped in a first reading.

This text is a course in quantum physics and should be understood as such. In particular, the text is definitely dense: notions appear one after another. Moreover, some of the notions are objectively difficult. Even if we have made an effort to explain as much as possible, it may take time to assimilate them.

The best way to assimilate notions is to put them into practice. This is why several exercises are proposed and their full solution is given at the end of each chapter. Even if you have the feeling of not having understood everything, try to do the exercises, and if you don't manage to solve an exercise, read the solution, then come back and try again.

Additional material

The additional material comprises the sections "The Broader View" and "References and Further Reading" of chapters 1-7 and the whole of chapters 8 and 9.

These texts are normally not self-contained. For beginners, they are meant to suggest perspectives for further studies. They are also meant for teachers and other knowledgeable readers, in order to fill gaps between our presentation, which is quite original, and more traditional ones.

Specifically about "References and Further Reading":

- A short list of suggested readings is provided. In a book like ours, there is no point in trying to be exhaustive: a long list would rather confuse the reader, and for sure some meaningful references would be left out anyway. Of course, the reader may start from those references to find others that they find more suitable, or start directly with something else, found elsewhere.
- For those who would like to embark on some specific project: we strongly advise to study *experiments* that can be understood with the help of the theory provided in this text; our bibliography always goes in this direction.

PART 1
Notions and Formalism

Chapter 1

Introducing Quantum Physics with Polarization

1.1 Polarization of a Light Beam

1.1.1 *Definition and basic measurement*

Light consists of electric and magnetic fields that can oscillate in any direction perpendicular to the direction of propagation. The polarization describes the direction of the electric field's oscillation. To measure this, we can use a polarizer, which is a material with a preferential axis due to its molecular or crystalline structure. This preferential axis allows the polarizer to act as a filter, transmitting only light polarized in certain directions.

We can liken polarized light being transmitted through a polarizer to metal bars with different orientations passing through a narrow door, which only allows vertically-oriented metal bars to pass through. Similarly, only light with certain polarization can be transmitted through a polarizer.

This analogy has limitations however, because only metal bars that are parallel to the door can pass through it, while those that are tilted from the vertical cannot. In comparison, polarization can be described as vectors, and even if the electric field is oriented at an angle to the polarizer's axis, the component parallel to the axis can still be transmitted, while the perpendicular component is reflected or dissipated as heat.

Imagine a polarizer with its axis oriented in the vertical direction. You send light through it, with an intensity I and polarization at an angle α to the vertical. Let the transmitted component be I_T, and the reflected component be I_R. I_T is oriented parallel to the vertical axis of the polarizer, whereas I_R is horizontal, perpendicular to the axis.

As intensity is a measure of energy, the law of conservation of energy give us the relation

$$I = I_T + I_R\,. \tag{1.1}$$

Next, imagine doubling the intensity I of the light that enters the polarizer. Since the initial intensity I is arbitrary, if it is doubled, the ratio of the transmitted and reflected intensities should still be the same. Hence I_T and I_R are also doubled. By intuition, the new relation should then be

$$2I = 2I_T + 2I_R\,. \tag{1.2}$$

We observe that $I_T \propto I$, and similarly $I_R \propto I$. Thus we obtain the relations

$$\begin{aligned} I_T &= k_1 I\,, \\ I_R &= k_2 I\,, \end{aligned} \tag{1.3}$$

where k_1, k_2 are constants. Substituting Equation (1.3) into Equation (1.1), we can deduce that

$$k_1 + k_2 = 1\,. \tag{1.4}$$

Two positive numbers such that $k_1 + k_2 = 1$ can always be written as $k_1 = \cos^2\alpha$ and $k_2 = \sin^2\alpha$, so finally

$$\begin{aligned} I_T &= I\cos^2\alpha\,, \\ I_R &= I\sin^2\alpha\,. \end{aligned} \tag{1.5}$$

This result is also known as Malus' Law in classical electromagnetic theory.

1.1.2 *Series of polarizers*

Let us consider a series of two polarizers, both with horizontal polarization axes (Figure 1.1). If the intensity transmitted by the first is I_{T1}, what is the intensity I_{T2} transmitted by the second? The answer is, of course, $I_{T2} = I_{T1}$, assuming that there is no energy loss when the light passes through the second polarizer.

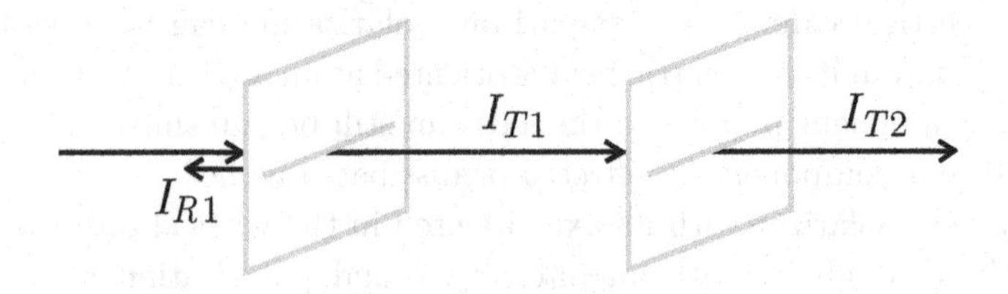

Fig. 1.1 Two polarizers with same polarization axes.

However, what happens if the second polarizer is tilted such that its polarization axis is now vertical (Figure 1.2)? The intensity I_{T2} now becomes zero.

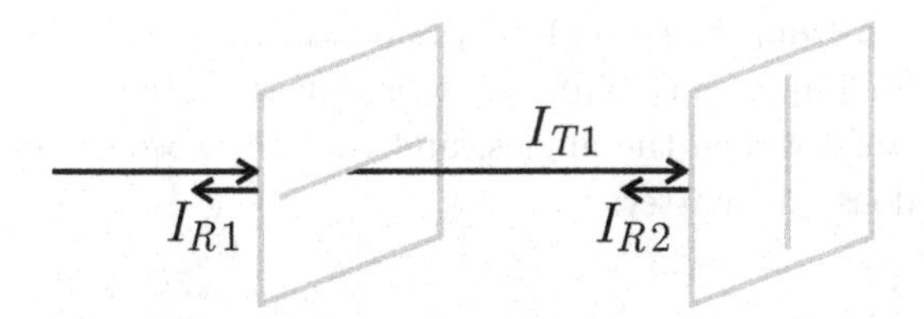

Fig. 1.2 Two polarizers with perpendicular polarization axes.

What happens next if a third polarizer is inserted between the two (Figure 1.3)? We expect that the intensity transmitted can decrease or remain the same if we add more polarizers, because more light would be filtered out, thus there should never be an increase. The curious thing is that, if this middle polarizer is oriented at a different angle from the other two, some light would pass through: the intensity actually increases! For example, if the middle polarizer has an axis oriented $\frac{\pi}{4}$ from the horizontal, the intensity transmitted by it would be $\frac{1}{2}I_{T1}$, and the intensity transmitted by the third polarizer would be half that from the second, that is, $\frac{1}{4}I_{T1}$.

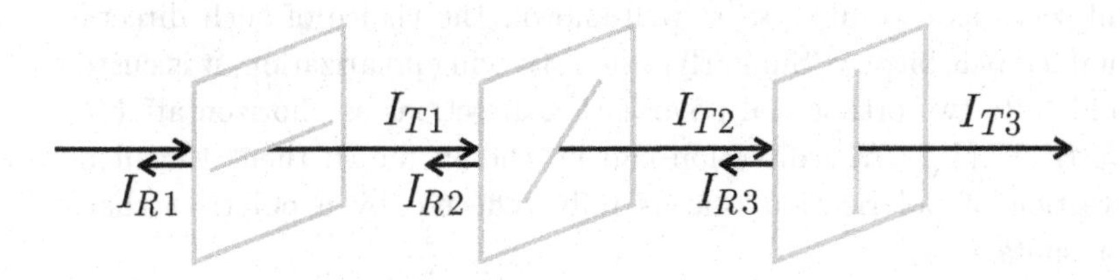

Fig. 1.3 Three polarizers each with polarization axis oriented $\frac{\pi}{4}$ with respect to the previous.

Hence polarizers are not merely filters that block a certain amount of light. This is a prediction of the laws and equations derived earlier, and it can easily be checked in a laboratory. The fact that this phenomenon has a surprising character when first encountered shows that our intuition sometimes fails, and we have to rely on observations and confirmed laws.

Now consider a case where we have the second and third polarizers oriented at angles α and β respectively from the first. How does the intensity transmitted by the third polarizer depend on α and β? From Equation (1.5), we know that the second polarizer will transmit a fraction $\cos^2\alpha$ of the intensity I_{T1} transmitted by the first. Since the third polarizer is oriented at $\beta - \alpha$ from the second, it would transmit a fraction $\cos^2(\beta - \alpha)$ of I_{T2}. Here, the important point to note is that the intensities depend only on the *difference* between the angles, and that the absolute orientation of the polarizer does not matter.

Exercise 1.1. A beam of horizontally-polarized light of intensity I impinges on a setup consisting of N polarizers, where N is very large. The first polarizer is oriented at an angle $\epsilon = \frac{\pi}{2N}$ to the horizontal; the next one is at an angle ϵ from the previous, and so on until the final one, which is exactly vertical. What is the intensity of light at the output of the last polarizer? What is its polarization? Note: neglect multiple reflections of the reflected beams and study only the transmitted beam.

1.1.3 *Polarization in vector notation*

In mechanics, it is very convenient to introduce a set of axes $(\hat{x}, \hat{y}, \hat{z})$ that will serve as coordinates; as well-known, the choice of such directions is completely arbitrary. Similarly, when studying polarization, it is customary to identify two orthogonal polarization directions as "horizontal" (H) and "vertical" (V). Any direction can be chosen for H; then, V will be the direction of polarization that is fully reflected by a polarizer that fully transmits H.

As usual, a direction can be described by a *vector*. Let us label the unit vectors along the vertical and horizontal directions as $\widehat{e}_V$ and $\widehat{e}_H$ respectively. Note that polarization is defined by the line along which the electric field oscillates, not by the direction along which the field points at a given time. Therefore, $-\widehat{e}_H$ describes the same polarization as $\widehat{e}_H$, and similarly for $\widehat{e}_V$.

With this basis of two vectors, one can describe any possible polarization direction: if the electric field is oscillating in a direction that makes an angle α with H, its polarization is[1]

$$\widehat{e}_\alpha = \cos\alpha\,\widehat{e}_H + \sin\alpha\,\widehat{e}_V \tag{1.6}$$

[1]Note that for simplicity we shall restrict our discussion to linear polarization throughout the main text; circular and elliptic polarization can be treated in a similar way by allowing complex numbers.

The polarization orthogonal to $\widehat{e}_\alpha$, i.e. the one that is fully reflected by a polarizer that fully transmits $\widehat{e}_\alpha$, is

$$\widehat{e}_{\alpha+\frac{\pi}{2}} \equiv \widehat{e}_{\alpha^\perp} = -\sin\alpha\,\widehat{e}_H + \cos\alpha\,\widehat{e}_V\,. \tag{1.7}$$

The remark we made on the previous page about the sign, of course, applies to all polarization vectors: $-\widehat{e}_\alpha$ describes the same polarization as $\widehat{e}_\alpha$. Note however that relative signs do matter: $\widehat{e}_\alpha = \cos\alpha\,\widehat{e}_H + \sin\alpha\,\widehat{e}_V$ describes a different polarization as $\cos\alpha\,\widehat{e}_H - \sin\alpha\,\widehat{e}_V \equiv \widehat{e}_{-\alpha}$.

1.1.4 *Polarizing beam-splitters as measurement devices*

We presented below the basic measurement of polarization, using a polarizer. For what follows, it will be rather convenient to keep in mind another device that measures polarization, namely the *polarizing beam-splitter* (Figure 1.4). Like the polarizer, this device splits light into two beams according to its polarization: for instance, horizontal polarization is transmitted and vertical polarization is reflected. However, here the reflected beam is not back-propagating or absorbed, as is the case with polarizers: rather, it is deflected in a different direction. This makes it easy to monitor both beams.

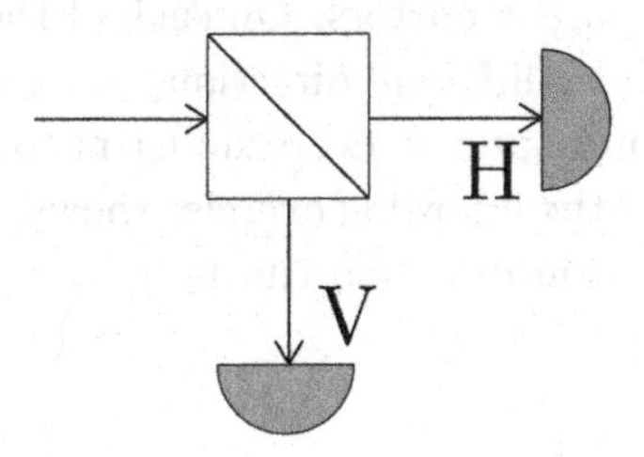

Fig. 1.4 Polarizing beam-splitter: scheme and implementation of a general measurement.

Note that one cannot rotate the device itself to measure along an angle α. In order to measure polarization along any arbitrary axis, polarizing beam-splitters must be complemented with *polarization rotators.* These are basically slabs of suitable materials, that can be chosen in order to perform the rotation

$$\widehat{e}_\alpha \longrightarrow \widehat{e}_H\,,$$
$$\widehat{e}_{\alpha^\perp} \longrightarrow \widehat{e}_V\,.$$

The measurement setup is now easy to understand: before the polarizing beam-splitter, one inserts the suitable rotator. If the output light is transmitted, then it was polarized along $\widehat{e}_H$ after the rotator, which means that it was polarized along $\widehat{e}_\alpha$ at the input. If on the contrary the output light is reflected, then it was polarized along $\widehat{e}_V$ after the rotator, which means that it was polarized along $\widehat{e}_{\alpha\perp}$ at the input.

1.1.4.1 *Reconstructing polarization: tomography*

We conclude this section with an important remark on measurement. We have described measurement schemes that only allow one polarization to be distinguished from the orthogonal one. One might ask if there exist measurements that would give more detailed information: for instance, one that would discriminate perfectly between three or more possible polarization directions.

On a single beam of light, this is not possible; the measurements that we described are optimal. However, if the beam of light is intense, a much more refined measurement of polarization can be made: one just has to first split the beam into several beams without affecting the polarization. This can be done with half-transparent mirrors. On each of the beams, polarization can be measured along a different direction. In particular, if the total intensity is known, two measurements are sufficient to completely determine a linear polarization[2], as the following exercise shows. Such a measurement, from which the polarization direction can be fully reconstructed, is called *tomography*.

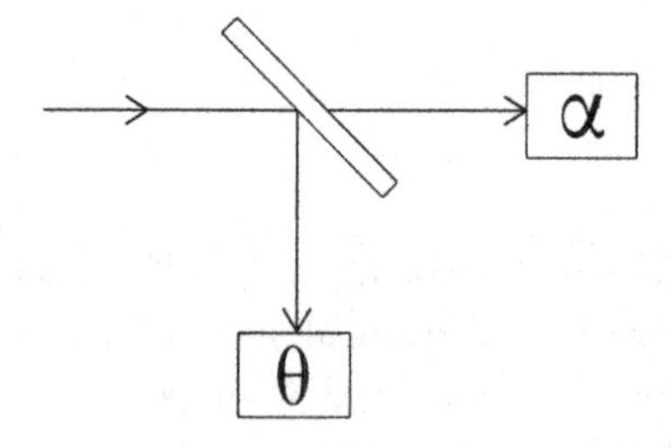

Fig. 1.5 Tomography of linear polarization.

[2] If the polarization is not guaranteed to be linear, three measurements would be needed.

Exercise 1.2. A beam of intensity $2I$ is split into two beams of intensity I each. On one of these beams, polarization is measured in the horizontal-vertical direction: the results are $I_H = I\cos^2\theta$, $I_V = I\sin^2\theta$. From these measurements, one can infer that the polarization is either $\widehat{e}_\theta = \cos\theta\,\widehat{e}_H + \sin\theta\,\widehat{e}_V$ or $\widehat{e}_{-\theta} = \cos\theta\,\widehat{e}_H - \sin\theta\,\widehat{e}_V$. Show that these two alternatives can be discriminated by measuring the polarization of the other beam in any other direction. Hint: suppose that the polarization is $\widehat{e}_\theta$: what are I_α and $I_{\alpha\perp}$, with $\widehat{e}_\alpha$, $\widehat{e}_{\alpha\perp}$ defined as in Equations (1.6), (1.7)? Compare these with the results obtained if the polarization is $\widehat{e}_{-\theta}$.

Using tomography, it seems that one can always learn the polarization direction perfectly. In classical physics, this is indeed the case. But the tomography process assumes that a light beam can always be split: what happens then, if the beam of light consists of the smallest unit, called a "photon"? This question marks the transition to quantum physics.

1.2 Polarization of One Photon

1.2.1 *Describing one photon*

During the development of modern physics, Einstein postulated that light is actually comprised of particles called photons, instead of the classical wave theory, to explain the photoelectric effect. The development of quantum physics clarified the notion of the photon. For the sake of this text, we can assume that light is "made" of photons.

In a polarized beam, each photon must have the same polarization. Indeed, if a polarizer is suitably oriented, the whole beam can be transmitted, that is each photon is transmitted.

Now, when dealing with a single photon, a different notation is used than when dealing with the whole beam. Namely, for a beam of polarization $\cos\alpha\,\widehat{e_H} + \sin\alpha\,\widehat{e_V}$, the state of polarization of each photon is written as $|\alpha\rangle = \cos\alpha|H\rangle + \sin\alpha|V\rangle$. This notation is called the *Dirac notation*, and we shall carefully study it now as we will be using it throughout the rest of the text.

1.2.2 *Computations using Dirac notation*

As mathematical objects, "$|.\rangle$" are vectors with very similar properties as the $\hat{e}$ used for the field.

In particular, let us write

$$\begin{aligned} |H\rangle &\equiv \begin{pmatrix} 1 \\ 0 \end{pmatrix}, \\ |V\rangle &\equiv \begin{pmatrix} 0 \\ 1 \end{pmatrix}. \end{aligned} \tag{1.8}$$

Then

$$|\alpha\rangle = \cos\alpha|H\rangle + \sin\alpha|V\rangle \equiv \begin{pmatrix} \cos\alpha \\ \sin\alpha \end{pmatrix}. \tag{1.9}$$

The coefficient $\cos\alpha$ is, as in normal vector algebra, the *scalar product* of $|\alpha\rangle$ and $|H\rangle$. We write this scalar product as

$$\langle\alpha|H\rangle = \cos\alpha\,. \tag{1.10}$$

Similarly,

$$\langle\alpha|V\rangle = \sin\alpha\,. \tag{1.11}$$

In particular, by construction

$$\langle H|H\rangle = 1\,,\ \langle V|V\rangle = 1\,,\ \langle H|V\rangle = 0\,, \tag{1.12}$$

that is, $\{|H\rangle, |V\rangle\}$ is a basis.

From what we have discussed earlier, we know that for a beam the quantity $\cos^2\alpha$ represents the fraction of the beam intensity that is transmitted by a filter. But what does such a quantity mean at the level of single photons? A photon is an indivisible particle: when arriving at the filter, it can either be transmitted or reflected. Given this, $\cos^2\alpha$ represents the *probability* that the photon is transmitted. Thus we can state that the probability of a photon passing through horizontal and vertical polarizers, given an initial polarization angle α, is

$$\begin{aligned} P(\text{finding } H \text{ given } \alpha) &= P(H|\alpha) = \cos^2\alpha = |\langle H|\alpha\rangle|^2\,, \\ P(\text{finding } V \text{ given } \alpha) &= P(V|\alpha) = \sin^2\alpha = |\langle V|\alpha\rangle|^2\,. \end{aligned}$$

Indeed these two numbers sum up to 1 as they should for probabilities. The rule is valid for two arbitrary polarization states $|\psi_1\rangle$ and $|\psi_2\rangle$:

$$P(\text{finding } \psi_1 \text{ given } \psi_2) = |\langle\psi_1|\psi_2\rangle|^2. \tag{1.13}$$

This last equation is called *Born's rule for probabilities.*

It is important here to note that $|\alpha\rangle = \cos\alpha|H\rangle + \sin\alpha|V\rangle$ does not mean that some photons are polarized as $|H\rangle$ and some as $|V\rangle$. Rather, it describes a new polarization state whereby all the photons are polarized as $|\alpha\rangle$, but when they are measured in the H/V basis, they have probabilities $\cos^2\alpha$ and $\sin^2\alpha$ of being transmitted.

The appearance of the notion of probability in physics requires a careful discussion. Before that, to consolidate our understanding, we propose two exercises.

Exercise 1.3.

(1) Prove that $\{|\alpha\rangle, |\alpha^\perp\rangle\}$ is a basis for all α, where

$$\begin{aligned} |\alpha\rangle &= \cos\alpha|H\rangle + \sin\alpha|V\rangle\,, \\ |\alpha^\perp\rangle &= \sin\alpha|H\rangle - \cos\alpha|V\rangle\,. \end{aligned} \tag{1.14}$$

(2) Let $|\beta\rangle = \cos\beta|H\rangle + \sin\beta|V\rangle$. Compute the probabilities $P(\alpha|\beta)$ and $P(\alpha^\perp|\beta)$.

As discussed earlier, the probabilities depend only on the difference between the angles α and β, and not on their individual values. In general, when light polarized at an angle α to the horizontal passes through a polarizer with an axis oriented β to the horizontal, the parameter for determining the probability of transmission is $\alpha - \beta$, and

$$\begin{aligned} P(T) &= \cos^2(\alpha - \beta)\,, \\ P(R) &= \sin^2(\alpha - \beta)\,. \end{aligned} \tag{1.15}$$

1.2.3 *The meaning of probabilities*

Let us discuss in greater detail the meaning of probabilities in measurements on single photons. Consider a few photons with polarization $|\alpha\rangle = \cos\alpha|H\rangle + \sin\alpha|V\rangle$, moving towards a polarizer oriented in the horizontal direction. Let us first predict the results of this experiment. Based on the earlier analysis, we can determine the transmitted and reflected intensities to be

$$\begin{aligned} P(T) &= \cos^2\alpha\,, \\ P(R) &= \sin^2\alpha\,. \end{aligned} \tag{1.16}$$

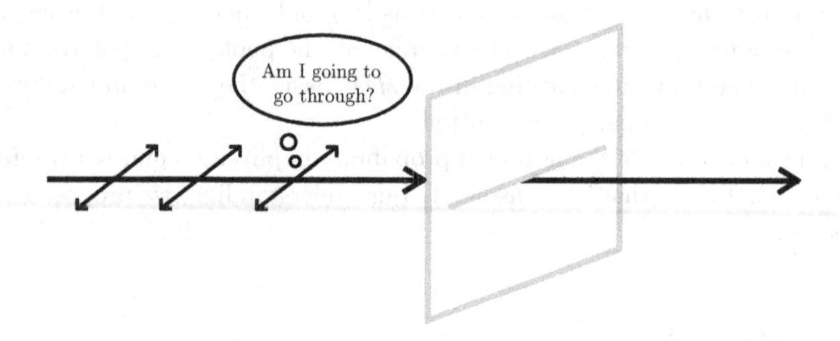

Fig. 1.6 Dilemma of a photon.

In other words, we can predict the statistical probability of the behaviour of a large number of photons, or a beam of light.

Now, let us consider each individual photon. Each photon is polarized along $|\alpha\rangle$ and is identical to each other. When one photon encounters the polarizer, what will happen to it? Will it pass through or will it be reflected? What about the next photon?

The answer is: we do not know. Although we know the statistical behaviour of many photons, we have no way of predicting the behaviour of each individual photon. Unsatisfied with not knowing the answer, we then ask, could it be that there is some hidden mechanism within each photon that would determine whether or not it would pass through?

The answer is, again, no. It has been experimentally proven that there is no such mechanism. In fact, we cannot predict the behaviour of each individual photon because the information is really not even there in the first place!

Our next question is, why can't this be predicted? Does this mean that nature is *random*? Does God play dice?

It may be surprising, but the answer is yes. The *randomness is intrinsic.* To make this point clearer, we can compare this situation to some well-known instances of randomness in our daily lives, such as tossing a coin, casting a die or weather patterns. In these cases, it may be hard or even impossible in practice to predict the results; however, this is not intrinsic randomness, since the relevant information is actually present.

The randomness in quantum phenomena is very different. We can measure the speed, polarization or any other parameters of a photon, but there is no way to tell whether or not it will pass through the polarizer. The information is really not there. This is what is meant by intrinsic randomness.

1.3 Describing Two Photons

Previously, in the case of a single photon, we were able to relate the statistical result of their intrinsically random behavior to the intensity of a light beam transmitted through a polarizer. However, two photons can exhibit purely quantum properties with no classical analogues.

1.3.1 *Classical composite systems*

In order to appreciate what follows, we consider first *composite systems* in classical physics. A physical system is composite if it can be seen as consisting of two or more subsystems. For example, the Earth and Moon. Neglecting its size, the physical properties of the Earth can be reduced to its position and speed $(\vec{x_E}, \vec{v_E})$; similarly for the Moon $(\vec{x_M}, \vec{v_M})$. Then the properties of the composite system can be derived from the set of vectors $(\vec{x_E}, \vec{v_E}, \vec{x_M}, \vec{v_M})$. The fact that the Earth and Moon interact means that, in time, the evolution of each parameter depends on the values of all the others. However, at any moment, the Earth has a well-defined position and speed; similarly for the Moon. This may seem obvious, and indeed it is in classical physics. But in quantum physics, this is not always the case: the properties of each subsystem may not be well-defined.

1.3.2 *Four states of two photons*

Consider now a composite system of two photons. Both photons could be horizontally polarized, or both vertically, or the first one horizontally and the second vertically, or the opposite. These situations are written as $|H\rangle \otimes |H\rangle$, $|V\rangle \otimes |V\rangle$, $|H\rangle \otimes |V\rangle$, and $|V\rangle \otimes |H\rangle$ respectively. The symbol $\otimes$ is a multiplication between vectors known as the tensor product defined as follows: let

$$|H\rangle = \begin{pmatrix} 1 \\ 0 \end{pmatrix}, \ |V\rangle = \begin{pmatrix} 0 \\ 1 \end{pmatrix}, \tag{1.17}$$

then

$$|H\rangle \otimes |H\rangle = \begin{pmatrix} 1 \\ 0 \end{pmatrix} \otimes \begin{pmatrix} 1 \\ 0 \end{pmatrix} = \begin{pmatrix} 1\begin{pmatrix} 1 \\ 0 \end{pmatrix} \\ 0\begin{pmatrix} 1 \\ 0 \end{pmatrix} \end{pmatrix} = \begin{pmatrix} 1 \\ 0 \\ 0 \\ 0 \end{pmatrix}, \tag{1.18}$$

and by using the same technique,

$$|H\rangle \otimes |V\rangle = \begin{pmatrix} 0 \\ 1 \\ 0 \\ 0 \end{pmatrix}, \quad |V\rangle \otimes |H\rangle = \begin{pmatrix} 0 \\ 0 \\ 1 \\ 0 \end{pmatrix}, \quad |V\rangle \otimes |V\rangle = \begin{pmatrix} 0 \\ 0 \\ 0 \\ 1 \end{pmatrix}. \tag{1.19}$$

Take note that this multiplication is not commutative, i.e $|H\rangle \otimes |V\rangle \neq |V\rangle \otimes |H\rangle$. This is obvious if you recall the physical meaning of this notation: "the first photon is H and the second is V" is definitely not the same as "the first photon is V and the second is H". Apart from this, the tensor product has all the usual properties of multiplication.

For simplicity, we shall write $|H\rangle \otimes |H\rangle$ as $|H, H\rangle$ or $|HH\rangle$, a convention that we will be using throughout this book.

1.3.3 *All the states of two photons*

Previously, we saw that the general expression for the linear polarization state of a single photon is $|\alpha\rangle = \cos\alpha|H\rangle + \sin\alpha|V\rangle$, and any α defines a valid state. Similarly, in the case of two photons, the most general state takes the form

$$|\psi\rangle = a|HH\rangle + b|HV\rangle + c|VH\rangle + d|VV\rangle . \tag{1.20}$$

To keep the interpretation of $|a|^2$, $|b|^2$, $|c|^2$ and $|d|^2$ as probabilities, we also impose the *normalization condition* $|a|^2 + |b|^2 + |c|^2 + |d|^2 = 1$. Conversely, any choice of $\{a, b, c, d\}$ satisfying the normalization condition represents a valid state.

Let us now study some examples. We start with

$$c|HH\rangle + s|HV\rangle = |H\rangle(c|H\rangle + s|V\rangle) , \tag{1.21}$$

where $c = \cos\alpha$ and $s = \sin\alpha$. This state represents the situation in which the first photon has a polarization of $|H\rangle$ while the second photon has a polarization of $|\alpha\rangle = c|H\rangle + s|V\rangle$.

Let us take a look at another state: $\frac{1}{\sqrt{2}}(|HH\rangle + |VV\rangle)$. Intuitively, this should also correspond to each photon having a well-defined polarization:

$$\frac{1}{\sqrt{2}}(|HH\rangle + |VV\rangle) \stackrel{?}{=} (c_\alpha|H\rangle + s_\alpha|V\rangle) \otimes (c_\beta|H\rangle + s_\beta|V\rangle) , \tag{1.22}$$

where $c_\alpha = \cos\alpha$ etc. Expanding the product, we get

$$\begin{aligned}\frac{1}{\sqrt{2}}|HH\rangle + \frac{1}{\sqrt{2}}|VV\rangle \stackrel{?}{=} c_\alpha c_\beta |HH\rangle + c_\alpha s_\beta |HV\rangle \\ + s_\alpha c_\beta |VH\rangle + s_\alpha s_\beta |VV\rangle\,.\end{aligned} \tag{1.23}$$

To satisfy this equality, we must have $c_\alpha c_\beta = s_\alpha s_\beta = \frac{1}{\sqrt{2}}$, but $c_\alpha s_\beta = s_\alpha c_\beta = 0$: the two conditions are manifestly contradictory. Hence, there is no way to write this state as the product of two independent states, one for each subsystem.

However, $\frac{1}{\sqrt{2}}(|HH\rangle + |VV\rangle)$ still describes a valid physical state, as we have mentioned earlier. We have to admit that there are some states for which it is not possible to describe the two photons separately. Such states are called *entangled*. It turns out that the conditions for composite states to be written as the product of two separate states are very constraining, and that vectors chosen at random are not likely to satisfy this. Thus most composite states are entangled.

Exercise 1.4. Consider the following two-photon states:

$$\begin{aligned}|\psi_1\rangle &= \frac{1}{2}(|HH\rangle + |HV\rangle + |VH\rangle + |VV\rangle)\,,\\ |\psi_2\rangle &= \frac{1}{2}(|HH\rangle + |HV\rangle + |VH\rangle - |VV\rangle)\,,\\ |\psi_3\rangle &= \frac{1}{2}|HH\rangle + \frac{\sqrt{3}}{2\sqrt{2}}(|VH\rangle + |VV\rangle)\,,\\ |\psi_4\rangle &= \cos\theta|HH\rangle + \sin\theta|VV\rangle\,.\end{aligned}$$

Verify that all the states are correctly normalized. Which ones are entangled? Write those that are *not* entangled as an explicit product of single-photon states.

1.3.4 *The meaning of entanglement*

What does entanglement mean? Let us stress explicitly that we are studying states, i.e. the description of the system, not its dynamics. The Earth and the Moon mentioned earlier influence the *motion* of each other, but we stressed (it seemed obvious) that at each time one can assign a state to the Earth and a state to the Moon. On the contrary, an entangled state describes a situation in which, at a given time and without any mention of evolution, two photons cannot be described separately. How should we understand this?

It is very hard, if not impossible, to give an intuitive meaning to entanglement since it is not observed in everyday life. However, physicists have performed tests on entangled photons and observed their properties. Some of these tests would be described in detail in the following chapters. We conclude with an intriguing exercise:

Exercise 1.5. Consider $|\alpha\rangle$ and $|\alpha^{\perp}\rangle$ as defined in Equation (1.14). Check that

$$\frac{1}{\sqrt{2}}\left(|\alpha\rangle|\alpha\rangle + |\alpha^{\perp}\rangle|\alpha^{\perp}\rangle\right) = \frac{1}{\sqrt{2}}\left(|H\rangle|H\rangle + |V\rangle|V\rangle\right). \quad (1.24)$$

What happens if both photons are measured in the same basis?

Notice that the polarization of the two photons are always the same, for all measurements. We say that their polarization are perfectly correlated. However, this correlation is not caused by each photon having the same well-defined state, for example, one photon is $|H\rangle$, the other is also $|H\rangle$, thus both always have the same polarization. Recall that for entangled photons, each photon cannot be described separately. This may be counter-intuitive, but even though both photons do not have well-defined properties, their properties are correlated such that their relative states are well-defined. Using the same example, this means that although we cannot say that either photon is $|H\rangle$, if one is measured to be $|H\rangle$, the other must definitely be $|H\rangle$ as well. Thus we see the presence of correlations in the properties of entangled photons.

1.4 Transformations of States

We have discussed how to measure states and predict the results of the measurements. Another important concept is the transformation of states, and the properties that such transformations obey.

First, we have to stress that the measurement process discussed thus far is an optimal one. In other words, there is no way in which two states can be distinguished better than by measuring them using polarizers or polarizing beam splitters. This is important because otherwise, Born's rule of probabilities would not be accurate as there are more optimal measurements that will produce different probabilities.

Secondly, it is an assumption of modern physics that transformations between macroscopic states are *reversible*. This means that if a state $|\phi\rangle$ can be transformed to $|\theta\rangle$, $|\theta\rangle$ can also be transformed to $|\phi\rangle$. An example of this is the rotation of a vector: it can be reversed by a rotation in the opposite direction. It turns out that all reversible transformations on a polarization vector are indeed related to rotations.

A theorem due to Wigner proves that in quantum physics, all transformations must be *linear* and have the property of *unitarity*:

- A transformation T is linear if, for all pairs of vectors $\{|\psi_1\rangle, |\psi_2\rangle\}$,
$$T(|\psi_1\rangle + |\psi_2\rangle) = T(|\psi_1\rangle) + T(|\psi_2\rangle)\,. \tag{1.25}$$
- A transformation U is unitary if it preserves the scalar product of any two vectors:
$$\langle\phi_1|\phi_2\rangle = \langle U(\phi_1)|U(\phi_2)\rangle\,. \tag{1.26}$$

The proof of linearity is the most difficult part of Wigner's theorem, here we sketch the reason behind the necessity of preserving the scalar product.

According to Born's rule, two states are perfectly distinguishable if the scalar product $\chi = 0$, and are the same if $\chi = 1$, indicating that χ is a measure of the "distinguishability" of the two states in an optimal measurement.

Now, let us analyze what would happen if χ is not preserved. If we suppose that there is a transformation such that χ decreases, this means that the states become more distinguishable, which contradicts the assumption that the measurement is optimal. Indeed, before performing the measurement, we could apply the transformation. If instead χ increases, the assumption of reversibility implies that there exists a reversed transformation in which χ decreases. Thus the only remaining option is that χ does not change during a reversible evolution.

1.5 Summary

The polarization of light describes the direction of the electric field's oscillation, and is written using vectors. Polarizers or polarizing beam-splitters are used to determine the polarization, by measuring the transmitted and reflected light intensities. For single photons, their polarization states are written using the Dirac notation, and Born's rule determines the probability of transmission or reflection through a measuring device. The behavior of single photons is intrinsically random; only the statistical behavior of a large number of photons can be predicted. Two photons can form entangled states in which each photon does not have a well-defined state, but are correlated such that their relative states are well-defined. Reversible transformations are also possible for states; these must be linear and unitary.

1.6 The Broader View

In this section, we present some of the concepts mentioned in this chapter in greater depth. These include vectors, probability and the degree of freedom.

1.6.1 *Vectors*

Vectors are a basic mathematical concept used in quantum theory. General vectors have similar properties as the geometric vectors encountered in elementary mathematics courses. The main difference is the notation: instead of the commonly used $\vec{v}$, in quantum mechanics we use the Dirac notation, $|v\rangle$. We also write the scalar product as $\langle v_1, v_2\rangle$ or $\langle v_1|v_2\rangle$ instead of $\vec{v_1}\cdot\vec{v_2}$.

Vectors can be expressed as a sum of component vectors, which is usually done using unit vectors that form a basis. For example,

$$\begin{pmatrix} a \\ b \end{pmatrix} = a\begin{pmatrix} 1 \\ 0 \end{pmatrix} + b\begin{pmatrix} 0 \\ 1 \end{pmatrix} . \tag{1.27}$$

In the Dirac notation, $\begin{pmatrix} a \\ b \end{pmatrix}$ is written as $|v\rangle$, and $\begin{pmatrix} 1 \\ 0 \end{pmatrix}$, $\begin{pmatrix} 0 \\ 1 \end{pmatrix}$ as $|e_1\rangle$ and $|e_2\rangle$ respectively. Then Equation (1.27) can be written as

$$|v\rangle = a|e_1\rangle + b|e_2\rangle . \tag{1.28}$$

Notice that a is actually the scalar product of $|v\rangle$ and $|e_1\rangle$, or $a = \langle e_1|v\rangle$. Likewise, $b = \langle e_2|v\rangle$. $|e_1\rangle$ and $|e_2\rangle$ also form an orthonormal basis, which means that these conditions are satisfied: $\langle e_1|e_1\rangle = 1$, $\langle e_2|e_2\rangle = 1$ and $\langle e_1|e_2\rangle = 0$. Geometrically, $|e_1\rangle$ and $|e_2\rangle$ are orthogonal to each other and normalized to one.

1.6.2 *Probability*

Many events in our daily lives are random, e.g. the results of tossing a coin or casting a dice. That is, their behaviour is unpredictable in the short term. However, these events may have a regular and predictable pattern in the long term. In the case of tossing a coin, we find that after many tries, half of the tosses produces heads and the other half produces tails. The *probability* is the proportion of times an event occurs over a large number of repetitions. Mathematically, we define the probability of an event A as

$$P(A) = \frac{n(A)}{n(S)}, \tag{1.29}$$

where $n(A)$ is the number of times event A occurs and $n(S)$ is the total number of events.

The probability $P(A)$ of any event A satisfies $0 \leq P(A) \leq 1$. By definition, if $P(A) = 1$, then event A will definitely happen and if $P(A) = 0$, then event A will definitely not happen.

Two events A and B are independent if the probability of one event happening does not affect the probability of the other event happening. For example, if you toss a coin and the result is a tail, the result of this toss does not affect the result of the next toss. Hence each toss is independent. If A and B are independent, then $P(A \cap B) = P(A) \times P(B)$, where $A \cap B$ denotes that both events A and B occur.

1.6.3 *Degree of freedom*

You may have encountered the phrase "degree of freedom" in chemistry or physics courses, in which mono-atomic particles are described as having three degrees of freedom while diatomic particles have five. In this case, each degree of freedom is a type of motion that the particles can undergo. Here in this book, this term will be used more generally to denote a physical variable that describes the system under study.

Within a system, there are often many variables, but not all the variables are of interest. For example, if we have identified a car as our system, the variables would include its position, velocity, and also its colour, temperature, material etc. If our aim is to study the motion of the car, then the colour of the car would be of little significance and can be ignored. In contrast, the position and speed of the car are of great importance; we call them the *degrees of freedom* of the car.

Let us take another system as an example. Suppose the physical system is an electron and our aim is to examine its motion. Again, the speed and position of the electron are its degrees of freedom. Its motion can be described by the equation $m\ddot{\vec{x}} = -e\vec{E}(x)$, where $\vec{E}(x)$ is the electric field in the region, and e is the charge of the electron.

However, if instead we want to take into account how the electric field changes as the electron moves, our system would then be composed of both the electron and electric field. We would then have to express their relationship using Maxwell's equations, and the analysis becomes more complicated. Thus as the system under consideration changes, the method of analysis also changes.

Before we end this discussion, we illustrate an important difference between quantum and classical systems through set theory. We first consider a classical system: in the example of the car, we can describe its degrees of freedom using a velocity-position graph, as in Figure 1.7.

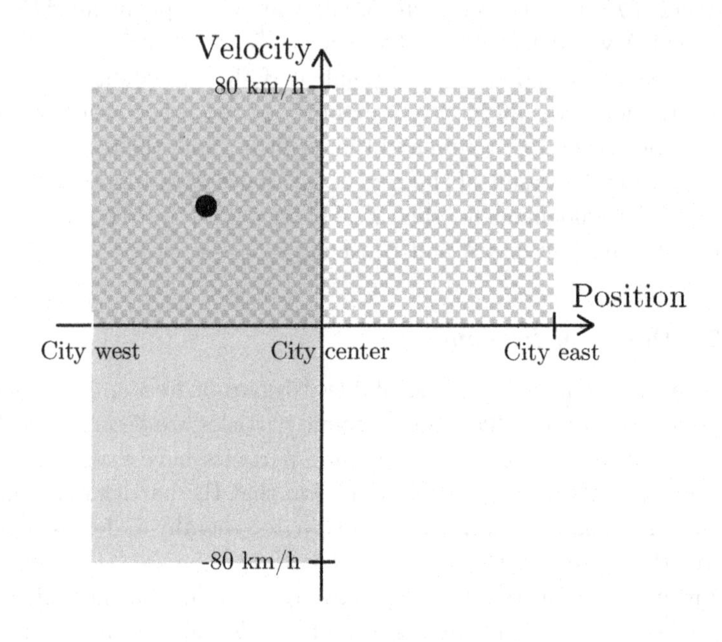

Fig. 1.7 Velocity-position graph.

The shaded region indicates all the cars that are between the west of the city and the city center, while the patterned region indicates all the cars that are traveling towards the east. By simple logic based on set theory, the overlapped region indicates the cars that are both between the west of the city and the city center, and are traveling towards the east. Thus if the car is at the position of the dot, we know that the car is between the west of the city and the city center, and is traveling at about 40 km/h towards the city center. We observe from this example that classical properties can be described using set theory.

To generalize this concept, suppose P_1 and P_2 are two physical properties, while S_1 and S_2 are sets that satisfy the physical properties P_1 and P_2 respectively. In mathematical notation,

$$\begin{aligned} P_1 &\to S_1 \,, \\ P_2 &\to S_2 \,. \end{aligned} \tag{1.30}$$

From what we have discussed thus far, we know that

$$\begin{aligned} &1.\, P_1 \wedge P_2 \;\to S_1 \cap S_2 \,, \\ &2.\, P_1 \vee P_2 \;\to S_1 \cup S_2 \,, \\ &3.\, P_1 \Rightarrow P_2 \to S_1 \subset S_2 \,. \end{aligned} \tag{1.31}$$

In words, the intersection of S_1 and S_2 satisfies the property P_1 and P_2, the union of S_1 and S_2 satisfies the property P_1 or P_2, and if S_1 is a subset of S_2, P_2 is satisfied if P_1 is satisfied.

However, this simple, intuitive relationship breaks down in quantum theory, because quantum systems do not have well-defined properties.

1.7 General References

Popular books on quantum physics:

- G.C. Ghirardi, *Sneaking a Look at God's Cards* (Princeton University Press, Princeton, 2003).
- V. Scarani, *Quantum Physics: A First Encounter* (Oxford University Press, Oxford, 2006).

General advanced textbooks:

- On classical optics: E. Hecht, *Optics* (Addison Wesley, San Francisco, 2002).
- A. Peres, *Quantum Theory: Concepts and Methods* (Kluwer, Dordrecht, 1995).
- M. Le Bellac, *A Short Introduction to Quantum Information and Quantum Computation* (Cambridge University Press, Cambridge, 2006).

1.8 Solutions to the Exercises

Solution 1.1. To solve this problem, first note that the intensity transmitted by each polarizer is a fraction $\cos^2\epsilon$ of the previous. Thus to find the intensity transmitted by the last one, we can multiply all these factors together:

$$\cos^2\epsilon \times \cos^2\epsilon \times ... = (\cos^2\epsilon)^N, \text{ where } N = \frac{\pi}{2\epsilon}.$$

We can then take the limit of the term $(\cos^2\epsilon)^N = (\cos^2\epsilon)^{\frac{\pi}{2\epsilon}}$ as ϵ approaches zero, and we find that

$$\lim_{\epsilon\to 0}(\cos\epsilon)^{\frac{\pi}{\epsilon}} = 1.$$

Thus we conclude that the intensity transmitted by the last polarizer is equal to the original intensity I! But the direction of polarization has changed: the light entered horizontally polarized and exited vertically polarized. In effect, this setup rotated the polarization while maintaining the intensity. A cultural remark: this effect, especially in the context of quantum physics, is called the *Zeno effect* from the name of a Greek philosopher; we let you guess why.

Solution 1.2. Suppose that the polarization is $\widehat{e}_\theta$, we can find I_α and $I_{\alpha^\perp}$ with reference to Equation (1.5), in which the difference in angles is now $\theta - \alpha$.

$$I_\alpha = I\cos^2(\theta-\alpha),$$
$$I_{\alpha^\perp} = I\sin^2(\theta-\alpha).$$

Similarly, if the polarization is $\widehat{e}_{-\theta}$, then I_α and $I_{\alpha^\perp}$ are given as follows:

$$I_\alpha = I\cos^2(\theta+\alpha),$$
$$I_{\alpha^\perp} = I\sin^2(\theta+\alpha).$$

This difference in the resulting I_α and $I_{\alpha^\perp}$ can be measured, thus the two polarization directions can be distinguished.

Solution 1.3.

(1) A basis is defined by

$$\langle\alpha|\alpha\rangle = \cos^2\alpha + \sin^2\alpha = 1,$$
$$\langle\alpha^\perp|\alpha^\perp\rangle = \cos^2\alpha + \sin^2\alpha = 1,$$
$$\langle\alpha|\alpha^\perp\rangle = \cos\alpha\sin\alpha - \cos\alpha\sin\alpha = 0.$$

These conditions can indeed be verified, either by writing both $|\alpha\rangle$ and $|\alpha^\perp\rangle$ as in Equation (1.9), or directly by using Equation (1.12).

(2) Using Born's rule for probabilities,

$$P(\alpha|\beta) = |\langle\alpha|\beta\rangle|^2 = |\cos\alpha\cos\beta + \sin\alpha\sin\beta|^2 = \cos^2(\alpha-\beta)\,,$$
$$P(\alpha^\perp|\beta) = |\langle\alpha^\perp|\beta\rangle|^2 = |\sin\alpha\cos\beta - \cos\alpha\sin\beta|^2 = \sin^2(\alpha-\beta)\,.$$

Solution 1.4. $|\psi_1\rangle$ is *not* entangled: it can be written as $\frac{1}{\sqrt{2}}(|H\rangle + |V\rangle) \otimes \frac{1}{\sqrt{2}}(|H\rangle + |V\rangle) = |\alpha = \frac{\pi}{4}\rangle \otimes |\beta = \frac{\pi}{4}\rangle$. All the other states are entangled (of course, $|\psi_4\rangle$ is not entangled if $\cos\theta = 0$ or $\sin\theta = 0$, but it is in all the other cases).

Solution 1.5.

$$\frac{1}{\sqrt{2}}\Big[|\alpha\rangle|\alpha\rangle + |\alpha^\perp\rangle|\alpha^\perp\rangle\Big]$$
$$= \frac{1}{\sqrt{2}}\Big[(c|H\rangle + s|V\rangle)(c|H\rangle + s|V\rangle) + (s|H\rangle - c|V\rangle)(s|H\rangle - c|V\rangle)\Big]$$
$$= \frac{1}{\sqrt{2}}\Big[(c^2+s^2)|HH\rangle + (cs - sc)|HV\rangle + (sc - cs)|VH\rangle + (s^2+c^2)|VV\rangle\Big]$$
$$= \frac{1}{\sqrt{2}}\Big[|HH\rangle + |VV\rangle\Big].$$

PART 2
Six Pieces

Chapter 2

Quantum Cryptography

Cryptography, as the name implies, is the art of sending cryptic messages for secure communication. The aim is to prevent any third party, Eve, from eavesdropping on the message between the sender and receiver, traditionally called Alice and Bob. Cryptography consists of two main steps: encoding and decoding. Typically, the sender encodes a plain message into a cryptic one which is sent to the receiver, who then decodes it to recover the original plain message. This procedure, a systematic way to encode and decode messages, is called a *cryptographic protocol*.

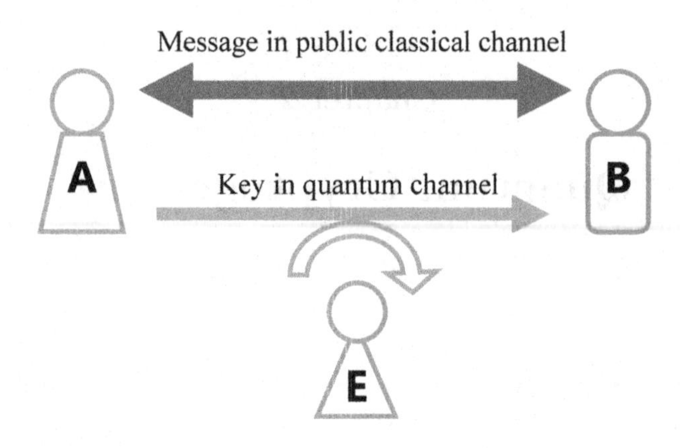

Fig. 2.1 Alice, Bob and Eve.

To date, there are two main classes of cryptographic protocols: the public and private key protocols. The *public key* protocol is used to transmit information from one sender to many receivers. These usually use the RSA algorithm, named after Rivest, Shamir and Adleman who devised it. This algorithm is based on the fact that it takes a lot of time to factorize a very large number into prime numbers.[1]

In the RSA algorithm, there is an assumption: that Eve has the same computational power as Alice and Bob (in short, her computers are not much more powerful than those available on the market). The second class of protocols, those that use a *private key*, are secure without this assumption. This class of protocols is used for communication between two people, who share a common secret code or *key*. The message is safe unless the eavesdropper knows this key. But there is a tradeoff for this security: the need to send the key without interception. How can this be done?

This is where quantum physics comes into play, opening up a new field called *quantum key distribution*. In this chapter, we first describe how private key cryptography works in practice by describing the one-time pad

[1]In this algorithm, the receiver, Bob, chooses two large prime numbers p and q. Bob then publicly reveals two related numbers: N, where $N = pq$ and c, a number having no common divisor with $(p-1)(q-1)$. Alice, the sender, uses N and c to encode the message that she wants to send. However, to decode the message, one has to know both p and q. Since it takes a lot of time to factorize N, if N is large, the message remains secret for a long time.

protocol. Then we study the first and most famous protocol for quantum cryptography.

2.1 One-Time Pad

Let us first prove that if Alice and Bob share a secret key, then they can communicate securely.

Suppose Alice wants to send Bob a greeting message "Hi". Here is the procedure for Alice to encode her message using the *one-time pad*:

(1) Alice needs to express the word "Hi" using bits. One way to do this is to use the ASCII scheme, in which each letter is assigned a sequence of 8 bits. This results in a message $\mathcal{M}$ of N bits. In this case, $N = 2*8 = 16$ bits.

Plain message	H	i
$\mathcal{M}$	01001000	01101001

(2) Let $\mathcal{K}$ be the secret key shared between Alice and Bob previously. Alice adds up $\mathcal{M}$ and $\mathcal{K}$ bit by bit, using a binary system in which $1+1=0$. This process creates a new list of random bits, $\mathcal{X}$. Mathematically, $\mathcal{X} = \mathcal{M} \oplus \mathcal{K}$, where $\oplus$ denotes the bitwise sum modulo 2.

Message $\mathcal{M}$	01001000	01101001
Key $\mathcal{K}$	11010101	10010101
Cryptic message $\mathcal{X}$	10011101	11111100

(3) Alice then sends $\mathcal{X}$ to Bob. It can be proven that Bob can extract the message using the key and no one else can obtain any information about it. This is the content of the following exercise.

Exercise 2.1. Prove the following statements:

(1) Knowing the key, the message can be retrieved through the operation $\mathcal{X} \oplus \mathcal{K} = \mathcal{M}$.
(2) The string $\mathcal{X}$, which Alice sends to Bob, is random, i.e., it does not contain any information on $\mathcal{M}$. Hint: take a bit whose value is 0 in $\mathcal{X}$: what is the value of the corresponding bit in $\mathcal{M}$?

We have just proven that the message can only be retrieved using the key. Note that the argument does not make any assumption on the computational power of the eavesdropper: even if Eve has an infinitely fast computer, she has nothing to compute!

This method is called the 'one-time pad' because the key can only be used once. If the key is used more than once, Eve can find some pattern in the messages and possibly deduce the key. This happened during the World War II, in which the Soviet Union sent coded messages using this method. However, they made the mistake of recycling their keys, thinking that their enemies would not notice. This proved to be deadly: after the U.S. gathered tons of correspondence from the Soviet Union, the mathematicians were able to uncover the key and eavesdrop on messages sent by the Soviet Union. This is the famous story of the Venona project. The lesson to be learned here is that it does not pay to be lazy; this protocol is safe only if Alice and Bob generate a new key for each communication.

2.2 The Idea of Quantum Key Distribution

We have just shown that if Alice and Bob share a secret key, they will be able to communicate securely. The problem now is to distribute this key in such a way that it is known only to Alice and Bob, and protected from any eavesdropper Eve.

Using classical methods of key distribution, such as via the Internet or telephone, it is impossible to know if Eve is eavesdropping on the communication channel. This is why in movies the characters have to meet and exchange suitcases: there is no way of distributing the key between distant partners... unless quantum physics is used. We have learnt in the first chapter that if a measurement is made on the polarization of a photon, its state of polarization would be changed. Thus if Eve eavesdrops on a channel sending polarized photons, the polarization states will be modified by her intervention. By checking if the result obtained by Bob corresponds to the polarization state sent by Alice, the presence of Eve can be detected. If no error is observed, it can be concluded that Alice and Bob share a secret key.

Before going into the details, it should be stressed again that quantum physics helps in distributing the key. Once the key is shared between Alice and Bob, the message will be sent using the classical one-time pad discussed above. Therefore, the precise name for quantum cryptography is *quantum key distribution*.

2.3 The BB84 Protocol

Let us now go into the details by outlining the protocol proposed by Bennett and Brassard in 1984, which is therefore known as BB84. This is the first quantum cryptography protocol to be proposed. Interestingly, physicists did not notice it because it was presented at a meeting of cryptographers! Quantum cryptography was later independently re-discovered by Ekert in 1991: this work was published in a physics journal and finally the idea was noticed.

2.3.1 *The steps of the protocol*

The steps are the following:

1. Quantum communication.

Alice sends linearly polarized photons to Bob. Each photon is prepared in one of these four states:

$$\begin{aligned} Z \text{ basis:} \quad |0_z\rangle &= |H\rangle\,, \\ |1_z\rangle &= |V\rangle\,. \\ X \text{ basis:} \quad |0_x\rangle &= |+\rangle = \frac{1}{\sqrt{2}}(|H\rangle + |V\rangle)\,, \\ |1_x\rangle &= |-\rangle = \frac{1}{\sqrt{2}}(|H\rangle - |V\rangle)\,. \end{aligned}$$

In other words, Alice uses either the Z basis (H/V) or the X basis ($+/-$), and in each basis, one of the states corresponds to the bit value 0 and the orthogonal one to the bit value 1.

Bob measures the polarization of each photon in either of these bases, choosing at random because he does not know the basis chosen by Alice.

For example, suppose Alice sends a $|0_z\rangle$ photon and Bob happens to orient his polarizer in the Z basis. The photon is transmitted through the polarizer and registers a count, thus Bob records the bit as 0. If Bob instead chooses to use the X basis, since $P(+|H) = P(-|H) = \frac{1}{2}$, Bob will register a 0 or a 1 with equal probability: there is no correlation between Alice's and Bob's bits when they use different bases.

2. Basis reconciliation.

At the end of the exchange of photons, Alice and Bob publicly reveal the basis in which each bit was coded or measured; the bits are kept if the same basis was used, and discarded if the bases were different. In this way, they remove the bits that have no correlation. Note that discarding half of the bits is not a problem since the key is a random list (while it would be detrimental to discard half of the bits in the message). The list of bits that are kept is called the *raw key.*

3. Classical post-processing.
Alice and Bob apply some classical information processing (*error correction*) to remove errors between their two lists. This procedure will tell them the amount of error present in their raw keys. If this number is small enough, they can apply another classical processing (*privacy amplification*) to make Eve's information negligible. If the error rate is too large, Alice and Bob discard the whole key. Two important remarks: first, privacy amplification is possible because, thanks to quantum physics, we can quantify Eve's information on the raw key. Second, if the amount of error in the raw key is too large, the raw key is discarded. A discarded key may seem to be a failure of the protocol since Eve could stop the exchange of the key by eavesdropping. But this protocol ensures that if a key is ever produced, it is safe.

In the above, notions such as the amount of error being too large and privacy amplification are vague. In the following, we will quantify and make these notions more rigorous.

2.3.2 *Statistics from quantum measurements*

If Eve is not present, i.e. no one is eavesdropping on the message, the raw key is already secret: neither error correction nor privacy amplification is needed. We study this case in the form of an exercise.

Exercise 2.2. There are a total of 16 possible measurement results for all the different combinations of bases chosen by Alice and Bob, since each of them can choose from 2 polarization bases. These are shown in the table on the next page. Fill up the last column, which is the probability of Bob obtaining the result in the second column if Alice sends the photon with polarization stated in the first column.

Alice sends	Bob measures ... and finds ...	Probability
$\|0_z\rangle$	$Z \to \|0_z\rangle$	1
$\|0_z\rangle$	$Z \to \|1_z\rangle$	0
$\|0_z\rangle$	$X \to \|0_x\rangle$	1/2
$\|0_z\rangle$	$X \to \|1_x\rangle$	1/2
$\|1_z\rangle$	$Z \to \|0_z\rangle$	
$\|1_z\rangle$	$Z \to \|1_z\rangle$	
$\|1_z\rangle$	$X \to \|0_x\rangle$	
$\|1_z\rangle$	$X \to \|1_x\rangle$	
$\|0_x\rangle$	$Z \to \|0_z\rangle$	
$\|0_x\rangle$	$Z \to \|1_z\rangle$	
$\|0_x\rangle$	$X \to \|0_x\rangle$	
$\|0_x\rangle$	$X \to \|1_x\rangle$	
$\|1_x\rangle$	$Z \to \|0_z\rangle$	
$\|1_x\rangle$	$Z \to \|1_z\rangle$	
$\|1_x\rangle$	$X \to \|0_x\rangle$	
$\|1_x\rangle$	$X \to \|1_x\rangle$	

With the help of this table, which describes an ideal situation without any error, verify that Alice and Bob indeed share a common secret key after basis-reconciliation.

We now consider the presence of Eve. We should consider the most general attack that Eve can do, but this requires knowledge that goes beyond the scope of this text. For the purpose of illustrating the principles, we present a specific attack called the *intercept-resend*. This attack is defined by having Eve do the same as Bob: she measures each photon sent by Alice, either in the Z or in the X basis. She then sends Bob a newly prepared photon, in the state corresponding to the result of her measurement.

For example, suppose Alice sends a photon $|0_z\rangle$ and Eve happens to measure it in the X basis. From the previous section, we know that Eve has a 0.5 probability of measuring either a 0 or a 1. Suppose Eve measures a 1, she would then send to Bob a photon $|1_x\rangle$. Now even if Bob chooses the same basis as Alice, namely Z, he may get the wrong result.

Now the situation is more complicated. Alice can choose between four polarization states, Eve and Bob can choose between the Z and X bases and measure either 1 or 0. We see that the total number of cases is $4*4*4 = 64$.

For simplicity, remember that in basis reconciliation, if Alice and Bob use different bases, the bits would be discarded regardless of Eve. Thus we can ignore these 32 cases. To shorten the list even further, we can consider the cases where Alice uses the Z basis only, the situation for the X basis being similar. Hence we only consider 16 cases in the exercise below, but bear in mind the other 16 cases in which Alice uses the X basis. Note the presence of errors: as noticed earlier, sometimes Alice and Bob don't have the same bit even though they used the same basis.

Exercise 2.3. Complete the table below: P_{Eve} is the probability of Eve measuring the polarization stated in the third column based on Alice's photon, while P_{Bob} is the probability of Bob measuring the polarization stated in the fifth column based on Eve's photon. P_{final} is the probability of Bob measuring the polarization stated in the fifth column based on Alice's photon.

No.	Alice sends	Eve	P_{Eve}	Bob	P_{Bob}	P_{final}
1	$\|0_z\rangle$	$Z \to \|0_z\rangle$	1	$Z \to \|0_z\rangle$	1	1
2	$\|0_z\rangle$	$Z \to \|0_z\rangle$	1	$Z \to \|1_z\rangle$	0	0
3	$\|0_z\rangle$	$Z \to \|1_z\rangle$		$Z \to \|0_z\rangle$		
4	$\|0_z\rangle$	$Z \to \|1_z\rangle$		$Z \to \|1_z\rangle$		
5	$\|0_z\rangle$	$X \to \|0_x\rangle$	1/2	$Z \to \|0_z\rangle$	1/2	1/4
6	$\|0_z\rangle$	$X \to \|0_x\rangle$		$Z \to \|1_z\rangle$		
7	$\|0_z\rangle$	$X \to \|1_x\rangle$		$Z \to \|0_z\rangle$		
8	$\|0_z\rangle$	$X \to \|1_x\rangle$		$Z \to \|1_z\rangle$		
9	$\|1_z\rangle$	$Z \to \|0_z\rangle$		$Z \to \|0_z\rangle$		
10	$\|1_z\rangle$	$Z \to \|0_z\rangle$		$Z \to \|1_z\rangle$		
11	$\|1_z\rangle$	$Z \to \|1_z\rangle$		$Z \to \|0_z\rangle$		
12	$\|1_z\rangle$	$Z \to \|1_z\rangle$		$Z \to \|1_z\rangle$		
13	$\|1_z\rangle$	$X \to \|0_x\rangle$		$Z \to \|0_z\rangle$		
14	$\|1_z\rangle$	$X \to \|0_x\rangle$		$Z \to \|1_z\rangle$		
15	$\|1_z\rangle$	$X \to \|1_x\rangle$		$Z \to \|0_z\rangle$		
16	$\|1_z\rangle$	$X \to \|1_x\rangle$		$Z \to \|1_z\rangle$		

At this point, you may answer the following questions:

(1) Why is it necessary to use two bases? What would happen if Alice and Bob decided to use just one basis to code the photons?
(2) How do Alice and Bob detect the presence of Eve?

2.3.3 *Extracting the secret key*

Having the statistics of the measurements, Alice and Bob can now apply error correction and privacy amplification: these will extract from the raw key a shorter key that is secret. But how much shorter? Again, a rigorous presentation requires information theory, which we have not studied here. However, the final result of those rigorous studies is intuitive and we can state it here.

Let I_{Bob} and I_{Eve} be a measure of Bob's and Eve's information on Alice's raw key respectively. Intuitively, if $I_{Bob} > I_{Eve}$, the situation is favorable and Bob can extract a secret key; on the contrary, if $I_{Bob} < I_{Eve}$, Eve would have too much information and the key would have to be discarded. Not only is this true, but for a suitable measure of information, the secret key will be shorter than the raw key by the factor

$$r = I_{Bob} - I_{Eve} \,. \tag{2.1}$$

In other words, if the raw key consists of N bits, the length of the secret key will be rN bits if $r > 0$ and 0 otherwise.

The following exercise proposes this analysis for the intercept-resend attack.

Exercise 2.4.

(1) I_{Eve} can be defined as the fraction of bits whose value Eve knows perfectly out of the total number of bits Alice sent, thus giving an estimate of Eve's average certainty of each photon. Referring to the table in the previous exercise, how much is I_{Eve} for the intercept-resend attack?

(2) The probability that Alice and Bob don't have the same bit, although they measured in the same basis, is called the *quantum bit error rate* (QBER, written as Q). Verify that $Q = 25\%$ for this attack.

(3) Accept without proof (see subsection 2.6.3) that Bob's information on Alice's string is given by the following formula:

$$I_{Bob} = 1 + Q \log_2 Q + (1 - Q) \log_2(1 - Q) \,, \tag{2.2}$$

with the convention that $0 \log_2 0 = 0$. Insert $Q = 25\%$ and compare I_{Bob} with I_{Eve}: can a secret key be extracted?

(4) Suppose now that Eve performs the intercept-resend attack with probability u, while with probability $1-u$ she lets the photon go to Bob untouched. Prove that in this case $Q = u \times 25\%$. How much is I_{Eve}?
(5) For which value of u does one have $I_{Bob} = I_{Eve}$? To what value of Q does it correspond?

2.4 More on Classical Post-Processing

The purpose of this section is to give an idea of the classical information processing that is performed in *error correction (EC)* and *privacy amplification (PA)*. Let us stress again that these procedures are entirely classical; they do not rely on any quantum formalism. However, privacy amplification requires knowledge of Eve's information, and there is no way of estimating this in classical cryptography. It is only in quantum cryptography that the knowledge of the error rate gives access to Eve's amount of information.

2.4.1 *Error correction*

We start at the point where Alice and Bob each have a raw key of N bits $\{a_i\}_{i=1\ldots N}$ and $\{b_i\}_{i=1\ldots N}$ and the two keys are not identical. Let the probability that the bits are equal be given by

$$p(a_i = b_i) = p > \frac{1}{2}\,. \tag{2.3}$$

The goal is to produce more correlated strings.

Here is a possible EC procedure: Alice and Bob take two successive bits (a_1, a_2) and (b_1, b_2), compute $A = a_1 \oplus a_2$ and $B = b_1 \oplus b_2$, and reveal the results. If $A = B$, they keep the first bit: $a_1' = a_1$, $b_1' = b_1$ and discard the other one; if $A \neq B$, they discard both bits. The following exercise shows how this procedure can indeed increase the correlations, at the expense of discarding more than half of the bits. In reality, much more efficient procedures are used that compare the parity of larger blocks and do not discard the whole block when an error is found.

Exercise 2.5. Consider the EC procedure described on the previous page:

(1) Preliminary question: why is "no information" given by $p = \frac{1}{2}$? Does the case $p = 0$ describe a situation which is favorable or unfavorable for Bob?

(2) Prove that the new strings are such that

$$p(a'_i = b'_i) = \frac{p^2}{p^2 + (1-p)^2} > p : \tag{2.4}$$

the probability that Alice's and Bob's bits have the same value has indeed increased.

(3) Of course, there is a price to pay: the strings have become shorter. Prove that the iteration decreases the length as

$$N \longrightarrow N' = \left[\frac{1}{2}\left(p^2 + (1-p)^2\right)\right] N . \tag{2.5}$$

(4) Suppose $p = 0.95$: how many iterations n are required to obtain $p^{(n)} = 0.99$? What is the length of the final string, as a function of the initial length N?

(5) Same question for an initial value $p = 0.6$.

(6) Why does one discard the second bit when $A = B$? (Suggestion: do not forget Eve).

2.4.2 *Privacy amplification*

Let's suppose now that Alice and Bob have corrected all their errors: $p(a_i = b_i) = 1$. Eve still has some information on Alice's bits:

$$p(e_i = a_i) = q > \frac{1}{2} . \tag{2.6}$$

In order to decrease Eve's information, Alice and Bob can apply the following PA procedure: as before, Alice and Bob start by taking two successive bits (a_1, a_2) and (b_1, b_2) and computing $A = a_1 \oplus a_2$ and $B = b_1 \oplus b_2$. Now, since their lists are equal, they certainly have $A = B$. Then, they set $a'_1 = A$, $b'_1 = B$. Eve wants to guess the new bit a'_1 by setting $e'_1 = e_1 \oplus e_2$. As the exercise shows, Eve's new guess is worse than her initial knowledge. Again, in practice, more elaborate procedures are used.

Exercise 2.6.

(1) Prove that

$$p(e_i' = a_i') = q^2 + (1-q)^2 < q : \tag{2.7}$$

the probability that Eve correctly guesses the bit has decreased. Again, the price to pay is that the strings have become shorter, by a factor of 2. Verify also that $q \to \frac{1}{2}$ after many iterations.

(2) Suppose $q = 0.95$. How many iterations are required to reach $q = 0.51$?

2.5 Summary

The basic principles of secure quantum cryptography involve the one-time pad and quantum key distribution. In the BB84 protocol, Alice sends the bit-encoded key via photons polarized in different bases, and Bob randomly chooses one of these bases for measurement. After this exchange of photons, Alice and Bob keep only the bits for which they used the same basis (basis reconciliation). Any intervention by Eve can be detected from errors in these bits. If this error rate is sufficiently small, classical post-processing (error correction and privacy amplification) can then be applied to establish a secret key.

2.6 The Broader View

2.6.1 *The power of Eve and the power of quantum*

Quantum cryptography has to deal with a very unusual standpoint. Normally, in physics, we are interested in describing what is actually done or observed. Quantum cryptography, on the contrary, has to describe *anything Eve could have done.* Eve's power is supposed to be limited only by the laws of physics. Surely, she cannot send a message faster than light, nor create a perfect copy of a quantum state (see the no-cloning theorem in the next chapter); but anything that is not forbidden in principle, she is allowed to do: she can have the fastest conceivable computer, she can perfectly entangle millions of photons... How far this can go from the simple intercept-resend attack that we described in the text!

The notions and mathematical tools needed to cope with such a new problem had to be invented: nowadays, they are available for many protocols. For instance, in the case of the BB84 implemented with single photons, it is known that Eve's largest possible information for an observed value of Q is

$$I_{Eve} = -Q \log_2 Q - (1-Q)\log_2(1-Q) = 1 - I_{Bob}\,. \qquad (2.8)$$

The secret key rate $I_{Bob} - I_{Eve}$ becomes zero for $Q > 11\%$ (compare with the result of the intercept-resend attack obtained above). Therefore, above this threshold, Eve might have used her power to perform a clever attack that compromises the key entirely: Alice and Bob should abort the protocol. But more importantly, we know now that a key *can* be extracted if $Q < 11\%$: if the error rate is below this threshold, whatever Eve might have done, Bob has more information than her on Alice's string.

2.6.2 *Practical quantum cryptography*

Since the typical Q observed in real experiments is well below 5%, the previous discussion might give the impression that everything is easy now: just run the BB84 protocol and extract a fully secret key. But it is not so simple.

First of all, all that we have done and mentioned in this introductory text is valid under the assumption that Alice's source produces perfect single photons. There is no such object, and most experiments rather use lasers that are very far from being single-photon sources. The whole security analysis must be reconsidered, and parameters other than the observed error rate start playing important roles (for instance, the intensity of the laser).

Then, when everything is well characterized on the quantum channel, one has to turn to Alice's and Bob's boxes and study them carefully. Of course, we have assumed that these boxes are private, but in a practical device are they really private? Take for instance the very first simple demonstration of QKD (it was not more sophisticated than what could be set up in the teaching lab of a school): there, when Bob changed his measurement basis, the devices made some noise. This gave rise to the famous pun that the setup was fully secure against a deaf eavesdropper! This anecdote is very silly and, by now, serious setups do not have such trivial flaws; but sometimes, someone finds more subtle ones. The ultimate check for security can probably never be made, one shall always have to live with some elements of trust.

2.6.3 *Information theory*

We have left one of the main equations in the text unexplained: Equation (2.2) which gives a formula for Bob's information on Alice's string of bits. This formula comes from classical *information theory.*

Information theory is an extremely broad and useful branch of applied mathematics. Started basically by Shannon in 1946, it is the basis of all modern communication and data processing. The first notion of this theory is the notion of *entropy*, basically a measure of "uncertainty before readout". The entropy of a biased coin, such that a head happens with probability p and a tail with probability $1-p$, is given by the quantity

$$H(\{p, 1-p\}) = -p \log_2 p - (1-p) \log_2(1-p) . \tag{2.9}$$

Now, notice that Equation (2.2) reads $I_{Bob} = 1 - H(\{Q, 1-Q\})$, i.e. one minus the uncertainty of Bob on Alice's bit due to the error Q: quite a reasonable definition for Bob's information.

We cannot delve here into the explanation of entropy's formula, its multiple roles in information theory or its link with entropy in thermodynamics: this will lead us far away from quantum physics, and anyway, fortunately there are many excellent references where these notions are discussed in depth.

If you are bored of quantum physics and want to try something else for your project or simply for your culture, information theory is certainly something to look into!

2.7 References and Further Reading

Easy reading:

- C.H. Bennett, G. Brassard, A. Ekert, Scientific American **267**, 50 (1992).

Resources:

- N. Gisin et al., Rev. Mod. Phys. **74**, 145 (2002).
- V. Scarani et al., Rev. Mod. Phys. **81**, 1301 (2009).

Suggestions for projects:
Experiments in quantum cryptography are usually very technical; we have selected two that are still reasonably simple here.

- W.T. Buttler et al., Phys. Rev. Lett. **81**, 3283 (1998).
- C. Kurtsiefer et al., Nature **419**, 450 (2002).

2.8 Solutions to the Exercises

Solution 2.1. To prove these statements, we start by constructing a table for the possible combinations of bits of $\mathcal{M}$ and $\mathcal{K}$.

Bits of $\mathcal{M}$	0	0	1	1
Bits of $\mathcal{K}$	0	1	0	1
$\mathcal{X} = \mathcal{M} \oplus \mathcal{K}$	0	1	1	0
$\mathcal{M} = \mathcal{X} \oplus \mathcal{K}$	0	0	1	1

From the last row, we see that Bob can obtain the original message $\mathcal{M}$ by performing the decoding operation $\mathcal{X} \oplus \mathcal{K}$.

We can also see from the table that if a bit in $\mathcal{M}$ is 0, the corresponding bit in $\mathcal{X}$ can be 0 or 1, and similarly for when $\mathcal{M}$ is 1. Thus we conclude that $\mathcal{X}$ does not contain any information on $\mathcal{M}$ if $\mathcal{K}$ is unknown.

Solution 2.2. Starting from the first row:

$$P(H|H) = 1\,,$$
$$P(V|H) = 0\,,$$
$$P(+|H) = \frac{1}{2}\,,$$
$$P(-|H) = \frac{1}{2}\,.$$

Using similar calculations, the table can be filled up as below:

Alice sends	Bob measures ... and finds ...	Probability
$\lvert 1_z\rangle$	$Z \to \lvert 0_z\rangle$	0
$\lvert 1_z\rangle$	$Z \to \lvert 1_z\rangle$	1
$\lvert 1_z\rangle$	$X \to \lvert 0_x\rangle$	1/2
$\lvert 1_z\rangle$	$X \to \lvert 1_x\rangle$	1/2
$\lvert 0_x\rangle$	$Z \to \lvert 0_z\rangle$	1/2
$\lvert 0_x\rangle$	$Z \to \lvert 1_z\rangle$	1/2
$\lvert 0_x\rangle$	$X \to \lvert 0_x\rangle$	1
$\lvert 0_x\rangle$	$X \to \lvert 1_x\rangle$	0
$\lvert 1_x\rangle$	$Z \to \lvert 0_z\rangle$	1/2
$\lvert 1_x\rangle$	$Z \to \lvert 1_z\rangle$	1/2
$\lvert 1_x\rangle$	$X \to \lvert 0_x\rangle$	0
$\lvert 1_x\rangle$	$X \to \lvert 1_x\rangle$	1

In basis reconciliation, measurements done using different bases are discarded. Hence all the measurements with probability $\frac{1}{2}$ would be discarded.

This leaves us with 8 combinations and we can easily see from the table that the bits measured by Bob correspond perfectly to the bits sent by Alice.

Solution 2.3. For the first row:

$$P_{Eve} = P(H|H) = 1\,,$$
$$P_{Bob} = P(H|H) = 1\,,$$
$$P_{final} = P_{Eve} * P_{Bob} = 1\,.$$

For the third row:

$$P_{Eve} = P(H|V) = 0\,.$$

Thus Eve would think that Alice's photon is $|0_z\rangle$, and would send this photon to Bob.

$$P_{Bob} = P(H|H) = 1\,,$$
$$P_{final} = P_{Eve} * P_{Bob} = 0\,.$$

For the fifth row:

$$P_{Eve} = P(+|H) = \frac{1}{2}\,,$$
$$P_{Bob} = P(H|+) = \frac{1}{2}\,,$$
$$P_{final} = P_{Eve} * P_{Bob} = \frac{1}{4}\,.$$

Use a similar logic to fill up the table.

No.	Alice sends	Eve	P_{Eve}	Bob	P_{Bob}	P_{final}
1	$\|0_z\rangle$	$Z \to \|0_z\rangle$	1	$Z \to \|0_z\rangle$	1	1
2	$\|0_z\rangle$	$Z \to \|0_z\rangle$	1	$Z \to \|1_z\rangle$	0	0
3	$\|0_z\rangle$	$Z \to \|1_z\rangle$	0	$Z \to \|0_z\rangle$	1	0
4	$\|0_z\rangle$	$Z \to \|1_z\rangle$	0	$Z \to \|1_z\rangle$	0	0
5	$\|0_z\rangle$	$X \to \|0_x\rangle$	1/2	$Z \to \|0_z\rangle$	1/2	1/4
6	$\|0_z\rangle$	$X \to \|0_x\rangle$	1/2	$Z \to \|1_z\rangle$	1/2	1/4
7	$\|0_z\rangle$	$X \to \|1_x\rangle$	1/2	$Z \to \|0_z\rangle$	1/2	1/4
8	$\|0_z\rangle$	$X \to \|1_x\rangle$	1/2	$Z \to \|1_z\rangle$	1/2	1/4
9	$\|1_z\rangle$	$Z \to \|0_z\rangle$	0	$Z \to \|0_z\rangle$	0	0
10	$\|1_z\rangle$	$Z \to \|0_z\rangle$	0	$Z \to \|1_z\rangle$	1	0
11	$\|1_z\rangle$	$Z \to \|1_z\rangle$	1	$Z \to \|0_z\rangle$	0	0
12	$\|1_z\rangle$	$Z \to \|1_z\rangle$	1	$Z \to \|1_z\rangle$	1	1
13	$\|1_z\rangle$	$X \to \|0_x\rangle$	1/2	$Z \to \|0_z\rangle$	1/2	1/4
14	$\|1_z\rangle$	$X \to \|0_x\rangle$	1/2	$Z \to \|1_z\rangle$	1/2	1/4
15	$\|1_z\rangle$	$X \to \|1_x\rangle$	1/2	$Z \to \|0_z\rangle$	1/2	1/4
16	$\|1_z\rangle$	$X \to \|1_x\rangle$	1/2	$Z \to \|1_z\rangle$	1/2	1/4

Note that errors occur when Eve uses a different basis from Alice.

(1) If they both use only a single basis, Eve can use the same basis to measure the bits sent by Alice, and would thus be able to determine all the bits. On the other hand, if two bases are used, Eve would only use the same basis as Alice half the time on average, thus she would only be able to accurately determine half of the entire key. Hence we observe that using two bases prevents Eve from knowing the key completely.
(2) If Alice publicly reveals some of her bits, Bob can compare his own measured ones with that of Alice. If the bits are different despite the same basis being used, then there is a probability that Eve had intercepted Alice and sent the wrong bit to Bob. For example, this case can be seen in row 6 of the table on the previous page.

Solution 2.4.

(1) $I_{Eve} = \frac{1}{2}$, since Eve has to use the same basis as Alice to be able to know her bit perfectly, and Eve chooses the correct basis half the time on average.
(2) Consider the case in which Alice sends $|0_z\rangle$ (the first 8 rows), the case for $|1_z\rangle$ being symmetrical. The last column, P_{final}, tells us the probability of each row to happen.

$$Q = \frac{\sum P(\text{not same bit})}{\sum P(\text{all cases})} = \frac{\sum P_{final} \text{ of rows 6 and 8}}{\sum P_{final} \text{ of rows 1 to 8}} = 25\%$$

(3) $I_{Bob} = 1 + \frac{1}{4}\,\log_2(\frac{1}{4}) + (1 - \frac{1}{4})\,\log_2(1 - \frac{1}{4}) \;= 0.189$
$I_{Bob} = 0.189 < I_{Eve} = 0.5$
Hence a secret key cannot be extracted by Bob.
(4) Errors only occur for the photons intercepted by Eve. Thus Q is equal to 25% of the proportion of intercepted photons, or

$$Q = u * 25\%\,.$$

As for Eve, from Part (1) she only knows half the bits that she intercepts. Thus

$$I_{Eve} = u * 50\%\,.$$

(5) If $I_{Bob} = I_{Eve}$, then from Equation (2.2)

$$1 + Q\,\log_2 Q + (1 - Q)\,\log_2(1 - Q) = I_{Eve}\,.$$

Substituting $Q = u \times 25\%$ and $I_{Eve} = u \times 50\%$, we can solve this to obtain

$$u = 0.682\,,\ Q = 17.1\%\,.$$

This means that if a secret key is to be extracted, the maximum probability u for Eve to perform the intercept-resend attack is 0.682, and the maximum Q that Bob measures is 17.1%.

Solution 2.5.

(1) If $p = \frac{1}{2}$, then there is an equal probability that the bits are wrong or correct. Thus Bob cannot determine whether any bit is more likely to be wrong or correct, and hence he gains no information from p.

If $p = 0$, the situation is favourable for Bob, since it means that all the bits are wrong, thus Bob can simply change all his bits to accurately determine Alice's bits.

(2) If $A = B$, then either $a_1 = b_1$ and $a_2 = b_2$, or $a_1 \neq b_1$ and $a_2 \neq b_2$. Thus

$$p(A = B) = p(a_{1,2} = b_{1,2}) + p(a_{1,2} \neq b_{1,2}) = p^2 + (1-p)^2\,,$$

$$p(a_i' = b_i') = \frac{p(a_i = b_i)}{p(A = B)} = \frac{p^2}{p^2 + (1-p)^2} > p,\ \text{since } p > \frac{1}{2}\,.$$

This means that the probability that Alice's and Bob's bits are the same has increased.

(3) The bits are only kept when $A = B$, and half of these bits are discarded. Thus the length decreases as

$$N' = \frac{1}{2}\,p(A = B)\,N = \left[\frac{1}{2}\left(p^2 + (1-p)^2\right)\right]N\,.$$

(4) We know that

$$p^{(n)} = \frac{\left(p^{(n-1)}\right)^2}{\left(p^{(n-1)}\right)^2 + \left(1 - p^{(n-1)}\right)^2}\,.$$

Substituting $p^{(0)} = 0.95$ and doing as many iterations as required, we find that $n = 1$ gives the required probability, since $p^{(1)} = 0.997 > 0.99$, and $N^{(1)} = 0.45\,N$.

(5) Repeating the steps in the previous part, we find that $n = 4$ gives the required probability, since $p^{(4)} = 0.998 > 0.99$, and $N^{(4)} = 0.0125\,N$.

(6) By announcing the results A and B, Eve would know whether the two bits are the same or different. This reduces the 4 possible combinations of two bits to two combinations, hence Eve gains information. However, in the procedure Alice and Bob discard the second bit, and use only the first bit, which, to Eve, is equally likely to be either 1 or 0. Even though Eve knows the sum of the two bits, this sum does not give her any information about the first bit. Thus by discarding the second bit, Alice and Bob ensure that Eve does not gain extra information when they announce their results.

Solution 2.6.

(1) If $e'_i = a'_i$, this means that either $e_1 = a_1$ and $e_2 = a_2$, or $e_1 \neq a_1$ and $e_2 \neq a_2$. Thus

$$p(e'_i = a'_i) = p(e_{1,2} = a_{1,2}) + p(e_{1,2} \neq a_{1,2}) = q^2 + (1-q)^2 < q\,.$$

Hence the probability that Eve's bit is correct is smaller.

To verify that $q \to \frac{1}{2}$ after many iterations, we first assume that q converges, i.e. $q^{(n+1)} = q^{(n)}$ when $n \to \infty$. Then

$$q^{(n+1)} = \left(q^{(n)}\right)^2 + \left(1 - q^{(n)}\right)^2 = q^{(n)}\,,$$

which we can solve to obtain $q^{(n)} = \frac{1}{2}$.

(2) We know that

$$q^{(n)} = \left(q^{(n-1)}\right)^2 + \left(1 - q^{(n-1)}\right)^2.$$

Substituting $q^{(0)} = 0.95$, we can do 6 iterations to obtain $q^{(6)} = 0.501 < 0.51$.

Quantum Cloning Pte Ltd
Mode | Swap | Symmetric | Asymmetric | "N" Copies | Perfect
Price | $0 | $10 | $5 | N x $8 | N/A

Chapter 3

Quantum Cloning

In the previous chapter, we saw that Eve is unsuccessful in avoiding detection because she causes errors when she measures in the wrong basis. For her, it would be nice to be able to make several perfect copies of each photon that Alice sent. Then she could measure a few copies and determine the exact polarization of the photon, while sending Bob one of the copied photons. In this way, she could avoid detection completely since all the photons Bob receives would be in the same polarization state as those that Alice sent.

Fortunately for Alice and Bob and unfortunately for Eve, such perfect cloning cannot be achieved. One can however strive to achieve the best possible result (optimal cloning). These notions are studied in this chapter.

3.1 The No-Cloning Theorem

A physicist receives a photon in an unknown polarization state $|\psi\rangle$ and would like to prepare a second photon in the same state. To do so, he prepares another photon in a fixed state $|R\rangle$: this state is like a blank piece of paper ready for the important information to be transferred. Now he would like to implement the following process:

$$|\psi\rangle \otimes |R\rangle \xrightarrow{?} |\psi\rangle \otimes |\psi\rangle\,. \tag{3.1}$$

The following exercise proves that such a perfect cloning cannot be done. This fact is called the *no-cloning theorem*.

Exercise 3.1.

(1) Assume that the transformation is possible for a basis $\{|H\rangle, |V\rangle\}$, i.e.

$$|H\rangle \otimes |R\rangle \longrightarrow |H\rangle \otimes |H\rangle\,,$$
$$|V\rangle \otimes |R\rangle \longrightarrow |V\rangle \otimes |V\rangle\,,$$

is possible. If perfect cloning for any arbitrary state is possible, how is the state $|+\rangle = \frac{1}{\sqrt{2}}(|H\rangle + |V\rangle)$ transformed? Using the rule that quantum transformations must be linear, introduced in section 1.4, prove that the transformation leads to a different result.

(2) Prove similarly that perfect cloning is impossible even with the help of an additional system that would play the role of a "machine":

$$|\psi\rangle \otimes |R\rangle \otimes |M\rangle \xrightarrow{?} |\psi\rangle \otimes |\psi\rangle \otimes |M(\psi)\rangle\,. \qquad (3.2)$$

Here, $|M\rangle$ is the initial state of the machine, $|M(\psi)\rangle$ is its final state (which may depend on $|\psi\rangle$).

We see that perfect cloning cannot be achieved. On the other extreme, the transformation $|\psi\rangle \otimes |R\rangle \longrightarrow |\psi\rangle \otimes |R\rangle$, in which the reference state $|R\rangle$ is not changed at all, is certainly possible since it amounts to doing nothing. Between these two extremes, there must be an optimal cloning procedure that transforms $|R\rangle$ to a state as close as can be achieved to the state $|\psi\rangle$, without greatly affecting the original state. To find this procedure, we first discuss what we can achieve through classical cloning, and we will see that quantum physics allows us to realize a better cloning procedure.

3.2 Trivial Cloning of Quantum States

Imagine that we have a photon with state $|\psi\rangle$ and a reference photon $|R\rangle$ in a randomly chosen state. We pass these two photons into a machine or black box that randomly gives us one of the photons that we put in. Each photon has a $\frac{1}{2}$ probability of emerging with state $|\psi\rangle$ and a $\frac{1}{2}$ probability of emerging with state $|R\rangle$. Thus we can say that the state $|\psi\rangle$ has been partially cloned onto the photon with original state $|R\rangle$, at the cost of losing the original photon.

The probability of obtaining the state $|\psi\rangle$ from the black box can be written as $P = \frac{1}{2}|\langle\psi|\psi\rangle|^2 + \frac{1}{2}|\langle\psi|R\rangle|^2$. We cannot compute this exactly given an arbitrary state $|R\rangle$; however, assuming that all possible polarization states $|\psi\rangle$ are possible, the quantity $|\langle\psi|R\rangle|^2$ will take all values between 0 and 1, and it can be proved that its average is $\frac{1}{2}$. Thus for this procedure, the probability of finding the right state at the output is, on average, $P = \frac{1}{2} + \frac{1}{2} * \frac{1}{2} = \frac{3}{4}$. This value is also called the *fidelity* of the cloning process.

The cloning procedure described here is so trivial that there is no real cloning involved. We did not put in any effort to clone the photon at all. Hence if we want an optimal cloning protocol, we must ensure that it is at least better than this trivial cloning.

3.3 Optimal Cloning of Quantum States

The optimal quantum cloner is called the *Bužek-Hillery cloner*, from the names of the people who first proposed it. It takes one photon and converts it into two photons with polarization similar to the original, using a third system as a machine. This transformation therefore acts on three systems: *the photon whose state one wants to copy (A), the photon onto which the state is copied (R), and the machine which can be just a third photon (M).* By convention, R and M are initially set as $|H\rangle$. In the computational basis, the transformation reads:

$$\begin{aligned} |H\rangle|H\rangle|H\rangle &\rightarrow \sqrt{\tfrac{2}{3}}\,|H\rangle|H\rangle|H\rangle + \sqrt{\tfrac{1}{6}}\Big(|H\rangle|V\rangle + |V\rangle|H\rangle\Big)|V\rangle \\ |V\rangle|H\rangle|H\rangle &\rightarrow \sqrt{\tfrac{2}{3}}\,|V\rangle|V\rangle|V\rangle + \sqrt{\tfrac{1}{6}}\Big(|V\rangle|H\rangle + |H\rangle|V\rangle\Big)|H\rangle\,. \end{aligned} \tag{3.3}$$

We can verify (see Exercise 3.2) that unlike perfect cloning, this transformation looks the same for any input state $|\psi\rangle = \cos\theta|H\rangle + \sin\theta|V\rangle$, where θ is an arbitrary constant:

$$|\psi\rangle|H\rangle|H\rangle \rightarrow \sqrt{\frac{2}{3}}\,|\psi\rangle|\psi\rangle|\psi\rangle + \sqrt{\frac{1}{6}}\Big(|\psi\rangle|\psi^{\perp}\rangle + |\psi^{\perp}\rangle|\psi\rangle\Big)|\psi^{\perp}\rangle\,. \tag{3.4}$$

where $|\psi^{\perp}\rangle = \sin\theta|H\rangle - \cos\theta|V\rangle$. The fidelity of this procedure is $\frac{5}{6}$, better than that of classical cloning.

Exercise 3.2.

(1) Write the result of the transformation on $|\psi\rangle|H\rangle|H\rangle$ using linearity and (3.3). *Hint:* $|\psi\rangle|H\rangle|H\rangle = \cos\theta|H\rangle|H\rangle|H\rangle + \sin\theta|V\rangle|H\rangle|H\rangle \rightarrow ...$

(2) Take (3.4), insert $|\psi\rangle = \cos\theta|H\rangle + \sin\theta|V\rangle$ and $|\psi^{\perp}\rangle = \sin\theta|H\rangle - \cos\theta|V\rangle$ and open the expression up. Verify that the result is the same as in part 1. *Hint:* the calculation is lengthy but straightforward.

(3) We have just shown that the transformation has the same form for all possible polarization states. Let us do a case study for $|\psi\rangle = |H\rangle$. After the transformation, the three photons are measured in the $H-V$ basis: using (3.3), complete the following table with the probabilities of all the possible outcomes:

Outcome	Probability
$\|H\rangle\|H\rangle\|H\rangle$	2/3
$\|H\rangle\|H\rangle\|V\rangle$	
$\|H\rangle\|V\rangle\|H\rangle$	
$\|H\rangle\|V\rangle\|V\rangle$	
$\|V\rangle\|H\rangle\|H\rangle$	
$\|V\rangle\|H\rangle\|V\rangle$	
$\|V\rangle\|V\rangle\|H\rangle$	
$\|V\rangle\|V\rangle\|V\rangle$	

(4) Verify that the probability of finding the first photon in the initial state $|H\rangle$ is $\frac{5}{6}$. This means that the state of the first photon has been modified by the transformation.

(5) Verify that the fidelity of cloning, or the probability of finding the second photon in the desired state $|H\rangle$ is also $\frac{5}{6}$.

3.4 Other Quantum Cloning Procedures

The Bužek-Hillery cloning procedure that we have studied is optimal among the procedures for $1 \rightarrow 2$ symmetric universal cloning. Let us explain the meaning of these words and the corresponding possible generalizations:

- Instead of taking one copy and trying to produce two, one can in general consider $N \to M$ cloning: starting from N copies and producing $M > N$ ones.
- The meaning of *symmetric* is that the output copies have the same fidelity. One can also consider *asymmetric* cloning, in which each copy may have a different fidelity. For the $1 \to 2$ case, the extreme cases of asymmetric cloning are doing nothing and swapping: then, one of the photons has perfect fidelity and the other has average fidelity $\frac{1}{2}$ (note that the trivial symmetric cloning we presented earlier is just the mixture of these two asymmetric strategies).
- The meaning of *universal* is that all the input states are copied equally well. One can consider *state-dependent* cloning, in which some states are cloned better than others. For instance, in cryptography, when Eve eavesdrops on the BB84 she knows that Alice and Bob are going to exchange only four states: it is those states she may want to copy, not all the others.

We cannot delve into this wealth of different procedures here. As an exercise, we consider $N \to M$ symmetric universal cloning.

Exercise 3.3. The optimal fidelity for $N \to M$ symmetric universal cloning has been found to be

$$F_{N\to M} = \frac{MN + M + N}{M(N+2)} \,. \tag{3.5}$$

(1) Verify that for $N = 1$ and $M = 2$ one finds $F = \frac{5}{6}$ as proven above.
(2) Fix $M = N + 1$: verify that $F_{N\to N+1}$ tends to 1 when N approaches infinity. Is this expected?
(3) Same question for $M = 2N$.

3.5 Stimulated Emission and Quantum Cloning

Consider the emission of a photon by the decay of an electron in an atom from an excited state to the ground state. In the absence of any other photon, we suppose that the photon is emitted with no preferred polarization (this is a crude assumption, wrong in general but used here to simplify the discussion). However, if a photon in state $|H\rangle$ is already present, one finds that:

(i) The probability p to have the new photon in the state $|V\rangle$ has not changed; this is called *spontaneous emission* and is always present;
(ii) The probability of emitting $|H\rangle$ is however increased to $2p$ (spontaneous and stimulated); one says that the emission of a new photon in state $|H\rangle$ is *stimulated* by the presence of another photon in state $|H\rangle$.

Of course, this is independent of the basis, because we have not defined anything that would single the H-V basis out. This effect, anticipated by Einstein in 1917, is the basis for the possibility of the laser. The fidelity of this process is $\frac{5}{6}$, as can be verified in the following exercise, which is the same as that of optimal cloning. In fact, the mechanism of spontaneous and stimulated emissions is a realization of optimal cloning.

Exercise 3.4.
(1) One sends a photon in state $|H\rangle$ to an atom in an excited state; two photons are found at the output. What is the probability of having the two in state $|H\rangle$? What is the probability of having one $|H\rangle$ and one $|V\rangle$? *Hint:* these are conditional probabilities: we consider only the case where a second photon has indeed been emitted.
(2) One picks one of the two photons at random and measures in the H-V basis: verify that the probability of finding it in state $|H\rangle$ is $\frac{5}{6}$.

3.6 Summary

The no-cloning theorem states that perfect cloning of quantum states is impossible. Classical procedures can only achieve a fidelity, or probability of finding the right state at the output, of $\frac{3}{4}$. Quantum physics allows a higher fidelity of $\frac{5}{6}$ in the optimal Bužek-Hillery cloner, as well as in spontaneous and stimulated emissions of photons.

3.7 The Broader View

3.7.1 *Relation with biological cloning*

Biological cloning is an operation on molecules (the DNA); molecules are definitely quantum objects. But then, why is perfect biological cloning possible, since quantum cloning is not?

In order to conclude that perfect cloning is impossible, it is not enough to notice that some information is encoded in quantum objects — some people may even tell you that everything, ultimately, is a quantum object. Rather, one should try and find out *how* information is encoded. In DNA, the information is coded in the *nature* of the molecules, not in their quantum states. Any amino-acid in the chain is either Adenine, Guanine, Cytosine, or Thymine: these are perfectly distinguishable objects, therefore there is no problem in reading the information and copying it. Note that, even for information coded in states, perfect cloning of a basis (i.e. of a set of perfectly distinguishable states) is possible.

3.7.2 *Relation with broadcasting in telecommunication*

Internet pages are downloaded identically by millions of people. Are the photons propagating in optical fibers not subjected to the no-cloning theorem?

The core of the answer is the same as for the previous one: in telecommunication, quantum objects are used to carry only classical information. For instance, a photon carries a "0" if it arrives early, a "1" if it arrives late in a given time bin.[1] Still, the no-cloning theorem plays a role in telecommunication, even if engineers give it a different name.

Indeed, when one downloads information from the internet, it is normal that the information from the server to the user has to travel a long distance. Now, even if many photons are sent and the optical fibers are of good quality, there is always a large amount of loss due to scattering: in order for it arrive at the end user, the signal must be amplified regularly. But an optical amplifier uses stimulated emission, and we have learned that stimulated emission is always accompanied by spontaneous emission.

Let us be more precise. An amplifying medium basically consists of atoms prepared in an excited state, ready to decay and emit photons.

[1]This coding, and more elaborated versions thereof, is called "time multiplexing".

When a pulse of light arrives, the probability of emitting photons in the same mode is enhanced: this is how optical amplification works. But of course, even when there is no pulse of light present, some atoms may decay spontaneously, thus emitting light at a time when there was none.

Telecom engineers know very well that they have to keep their signals well above the threshold of spontaneous emission: this nuisance is a direct consequence of the no-cloning theorem of quantum physics.

3.8 References and Further Reading

Resources:

- V. Scarani et al., Rev. Mod. Phys. **77**, 1225-1256 (2005).
- N.J. Cerf, J. Fiurášek, Progress in Optics, vol. 49, Edt. E. Wolf (Elsevier, 2006), p. 455.

Suggestions for projects:

- S. Fasel et al., Phys. Rev. Lett. **89**, 107901 (2002).
- W.T.M. Irvine et al., Phys. Rev. Lett. **92**, 047902 (2004).

3.9 Solutions to the Exercises

Solution 3.1.

(1) Assuming that the transformation is possible for $|\psi\rangle$ belonging to a basis, let us transform $|+\rangle = \frac{1}{\sqrt{2}}(|H\rangle + |V\rangle)$:

$$\begin{aligned} |+\rangle \otimes |R\rangle &= \tfrac{1}{\sqrt{2}}(|H\rangle \otimes |R\rangle + |V\rangle \otimes |R\rangle) \\ &\longrightarrow \tfrac{1}{\sqrt{2}}(|H\rangle \otimes |H\rangle + |V\rangle \otimes |V\rangle) . \end{aligned} \tag{3.6}$$

If the state $|+\rangle$ is perfectly cloned, the resulting state should be

$$\begin{aligned} |+\rangle \otimes |+\rangle &= \tfrac{1}{\sqrt{2}}(|H\rangle + |V\rangle) \otimes \tfrac{1}{\sqrt{2}}(|H\rangle + |V\rangle) \\ &= \tfrac{1}{2}(|HH\rangle + |HV\rangle + |VH\rangle + |VV\rangle) , \end{aligned} \tag{3.7}$$

which is not the same as (3.6). Hence, we have proved that there is no transformation that can perfectly clone photons.

(2) Similar to the previous question, we first assume that the transformation is possible for $|\psi\rangle$ belonging to a basis:

$$|H\rangle \otimes |R\rangle \otimes |M\rangle \longrightarrow |H\rangle \otimes |H\rangle \otimes |M(H)\rangle\,,$$
$$|V\rangle \otimes |R\rangle \otimes |M\rangle \longrightarrow |V\rangle \otimes |V\rangle \otimes |M(V)\rangle\,.$$

Now, let us transform $|\psi\rangle = |+\rangle$,

$$|+\rangle|R\rangle|M\rangle \longrightarrow \frac{1}{\sqrt{2}}(|H\rangle|H\rangle|M(H)\rangle + |V\rangle|V\rangle|M(V)\rangle)\,. \tag{3.8}$$

We shall evaluate $|+\rangle \otimes |+\rangle \otimes |M\rangle$ for comparison:

$$|+\rangle|+\rangle|M\rangle = \frac{1}{2}(|HH\rangle + |HV\rangle + |VH\rangle + |VV\rangle) \otimes |M(+)\rangle\,. \tag{3.9}$$

Comparing (3.8) and (3.9), we see that the only way $|M(H)\rangle$, $|M(V)\rangle$ and $|M(+)\rangle$ can satisfy these equalities is for them to be 0. We hence conclude that perfect copying is impossible even with the help of an auxiliary system.

Solution 3.2.

(1) Using the clue,

$$\begin{aligned}
|\psi\rangle|H\rangle|H\rangle &= \cos\theta|H\rangle|H\rangle|H\rangle \;+\; \sin\theta|V\rangle|H\rangle|H\rangle \\
&\to \cos\theta\left(\sqrt{\frac{2}{3}}|HHH\rangle \;+\; \sqrt{\frac{1}{6}}\Big(|HV\rangle + |VH\rangle\Big)|V\rangle\right) \\
&\quad + \sin\theta\left(\sqrt{\frac{2}{3}}|VVV\rangle \;+\; \sqrt{\frac{1}{6}}\Big(|VH\rangle + |HV\rangle\Big)|H\rangle\right) \\
&= \sqrt{\frac{2}{3}}\cos\theta|HHH\rangle + \sqrt{\frac{1}{6}}\cos\theta\Big(|HVV\rangle + |VHV\rangle\Big) \\
&\quad + \sqrt{\frac{2}{3}}\sin\theta|VVV\rangle + \sqrt{\frac{1}{6}}\sin\theta\Big(|VHH\rangle + |HVH\rangle\Big)\,.
\end{aligned}$$

(2) To simplify the presentation, we let s and c stand for $\sin\theta$ and $\cos\theta$ respectively. To do this question, we need to use the trigonometric identity $\sin^2\theta + \cos^2\theta = 1$.

$$
\begin{aligned}
|\psi\rangle|H\rangle|H\rangle \to & \sqrt{\frac{2}{3}}|\psi\rangle|\psi\rangle|\psi\rangle + \sqrt{\frac{1}{6}}\Big(|\psi\rangle|\psi^\perp\rangle + |\psi^\perp\rangle|\psi\rangle\Big)|\psi^\perp\rangle \\
= & \sqrt{\frac{2}{3}}\Big(c|H\rangle + s|V\rangle\Big)\Big(c|H\rangle + s|V\rangle\Big)\Big(c|H\rangle + s|V\rangle\Big) \\
& + \sqrt{\frac{1}{6}}\Big(c|H\rangle + s|V\rangle\Big)\Big(s|H\rangle - c|V\rangle\Big)\Big(s|H\rangle - c|V\rangle\Big) \\
& + \sqrt{\frac{1}{6}}\Big(s|H\rangle - c|V\rangle\Big)\Big(c|H\rangle + s|V\rangle\Big)\Big(s|H\rangle - c|V\rangle\Big) \\
= & \sqrt{\frac{2}{3}}\cos\theta|HHH\rangle + \sqrt{\frac{1}{6}}\cos\theta\Big(|HVV\rangle + |VHV\rangle\Big) \\
& + \sqrt{\frac{2}{3}}\sin\theta|VVV\rangle + \sqrt{\frac{1}{6}}\sin\theta\Big(|VHH\rangle + |HVH\rangle\Big).
\end{aligned}
$$

(3) From Equation (3.3), we obtain the following table

Number	Outcome	Probability
1	$\|H\rangle\|H\rangle\|H\rangle$	2/3
2	$\|H\rangle\|H\rangle\|V\rangle$	0
3	$\|H\rangle\|V\rangle\|H\rangle$	0
4	$\|H\rangle\|V\rangle\|V\rangle$	1/6
5	$\|V\rangle\|H\rangle\|H\rangle$	0
6	$\|V\rangle\|H\rangle\|V\rangle$	1/6
7	$\|V\rangle\|V\rangle\|H\rangle$	0
8	$\|V\rangle\|V\rangle\|V\rangle$	0

(4) In the "Outcome" column above, looking at the first of the three photons in each state, we add up the probabilities for all outcomes with the first state as $|H\rangle$ (numbers 1 to 4):

$$P = \frac{2}{3} + \frac{1}{6} = \frac{5}{6}.$$

(5) Similiar to the previous question, we add up the probabilities of all outcomes with the second photon as $|H\rangle$ (numbers 1, 2, 5 and 6):

$$P = \frac{2}{3} + \frac{1}{6} = \frac{5}{6}.$$

Solution 3.3.

(1) Substituting the values for N and M into (3.5), we get

$$F_{1\to 2} = \frac{2+2+1}{2(1+2)} = \frac{5}{6}\,.$$

(2) Substituting $M = N+1$ and taking the limit as $N \to \infty$,

$$\lim_{N\to\infty} F_{N\to N+1} = \lim_{N\to\infty} \frac{N(N+1)+(N+1)+N}{(N+1)(N+2)} = 1\,.$$

(3) Substituting $M = 2N$ and taking the limit as $N \to \infty$,

$$\lim_{N\to\infty} F_{N\to 2N} = \lim_{N\to\infty} \frac{2N^2+2N+N}{2N^2+4N} = 1\,.$$

Solution 3.4.

(1) To have both photons in state $|H\rangle$, the new photon must be emitted in state $|H\rangle$; the probability of this is $2p$. Similarly, the probability of having one $|H\rangle$ and one $|V\rangle$ is p.
(2) If both photons are $|H\rangle$ (probability $2p$), picking one of these photons at random would definitely yield $|H\rangle$. On the other hand, if one is $|H\rangle$ and one is $|V\rangle$ (probability p), the probability of picking the photon in state $|H\rangle$ is half. Thus the total probability is

$$P = \frac{2p+\frac{1}{2}p}{2p+p} = \frac{5}{6}\,.$$

ALL SYSTEMS READY. QUANTUM TELEPORTATION TO MARS INITIATE!
ENTANGLED SOURCE A
TELEPORT
ENTANGLED SOURCE B
RECEIVING QUANTUM INFORMATION. READY FOR TELEPORTATION.
QUANTUM
CLASSICAL
TELEPORT

Chapter 4

Quantum Teleportation

Teleportation used to be only a fantasy of science fiction. However, quantum physics has achieved this seemingly inconceivable feat: physicists have developed a protocol that enables the teleportation of information from one place to another, without any direct interaction or transfer of mass. This means that Bob could obtain a photon with a certain polarization state from Alice without Alice actually describing the polarization or sending a copy of the photon over. How is this possible?

In this chapter, we shall examine the quantum teleportation protocol.

4.1 Teleportation Protocol

Let us start with a setup of three photons A, B and C. A is prepared by Alice in an arbitrary polarization state $|\psi\rangle = \cos\theta|H\rangle + \sin\theta|V\rangle$; B and C are in the entangled state $|\Phi^+\rangle = \frac{1}{\sqrt{2}}\left(|H\rangle|H\rangle + |V\rangle|V\rangle\right)$, and are sent to Alice and Bob respectively. Our goal now is to transfer the state of photon A to photon C using the set-up in Figure 4.1.

In a nutshell, the teleportation protocol works as follows: Alice takes photons A and B, and performs a measurement that entangles them (Bell measurement). The effect of this measurement is to somehow disentangle photons B and C. Moreover, photon C is at this point in a state related to the desired state $|\psi\rangle$. In order to retrieve $|\psi\rangle$ exactly, Alice has to send two bits of information to Bob, who then applies a unitary transformation to photon C. Let us go through this protocol in detail through the following exercises.

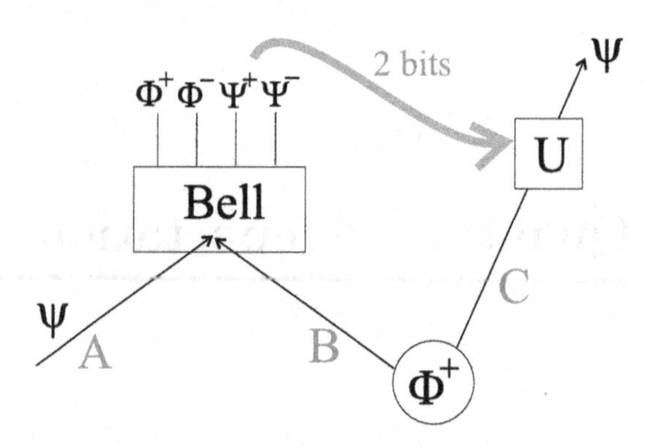

Fig. 4.1 Scheme for quantum teleportation.

Step 1: Entanglement

We decompose the initial state, in which B and C are entangled, in a basis of states in which A and B are entangled.

Exercise 4.1. Verify the following equality:

$$\begin{aligned} |\psi\rangle_A |\Phi^+\rangle_{BC} = & \ \tfrac{1}{2} |\Phi^+\rangle_{AB} \big(\cos\theta|H\rangle + \sin\theta|V\rangle\big)_C \\ & + \tfrac{1}{2} |\Phi^-\rangle_{AB} \big(\cos\theta|H\rangle - \sin\theta|V\rangle\big)_C \\ & + \tfrac{1}{2} |\Psi^+\rangle_{AB} \big(\cos\theta|V\rangle + \sin\theta|H\rangle\big)_C \\ & + \tfrac{1}{2} |\Psi^-\rangle_{AB} \big(\cos\theta|V\rangle - \sin\theta|H\rangle\big)_C , \end{aligned} \tag{4.1}$$

where the four states below are called Bell states.

$$\begin{aligned} |\Phi^+\rangle &= \tfrac{1}{\sqrt{2}} \left(|H\rangle|H\rangle + |V\rangle|V\rangle\right), \\ |\Phi^-\rangle &= \tfrac{1}{\sqrt{2}} \left(|H\rangle|H\rangle - |V\rangle|V\rangle\right), \\ |\Psi^+\rangle &= \tfrac{1}{\sqrt{2}} \left(|H\rangle|V\rangle + |V\rangle|H\rangle\right), \\ |\Psi^-\rangle &= \tfrac{1}{\sqrt{2}} \left(|H\rangle|V\rangle - |V\rangle|H\rangle\right). \end{aligned} \tag{4.2}$$

Step 2: Measurement

After photon B is sent to Alice, she performs a joint measurement of the polarization of photons A and B. This is called a Bell measurement, and the result of this measurement is one of the Bell states in Equation (4.2). Since photons B and C were initially entangled, the Bell measurement would also change the state of photon C.

Exercise 4.2. Fill up the table below:

Outcome of measurement A-B	Resulting state of C	Probability
$\|\Phi^+\rangle$	$\cos\theta\|H\rangle + \sin\theta\|V\rangle$	1/4
$\|\Phi^-\rangle$		
$\|\Psi^+\rangle$		
$\|\Psi^-\rangle$		

Step 3: Transformation

From the table above, we see that there are four possible resulting states of photon C after the Bell measurement on A-B. However, only one of these states, $\cos\theta|H\rangle + \sin\theta|V\rangle$, is the same as the original state of photon A, $|\psi\rangle$. To achieve teleportation, Bob thus has to perform a transformation on photon C. For example, if Alice measures $|\Phi^-\rangle$ on photons A-B, photon C would be in the state $\cos\theta|H\rangle - \sin\theta|V\rangle$. To obtain $|\psi\rangle$, Bob could perform the transformation $|H\rangle \to |T(H)\rangle = |H\rangle$, $|V\rangle \to |T(V)\rangle = -|V\rangle$ on C. For this transformation to be possible, it must be unitary, which means that the scalar product must be preserved.

Exercise 4.3.

(1) Verify that $\langle T(H)|T(H)\rangle = \langle H|H\rangle$, $\langle T(V)|T(V)\rangle = \langle V|V\rangle$ and $\langle T(H)|T(V)\rangle = \langle H|V\rangle$. This shows that the transformation is unitary and is hence possible.

(2) Now, find the transformations that bring $\cos\theta|V\rangle + \sin\theta|H\rangle$ and $\cos\theta|V\rangle - \sin\theta|H\rangle$ into $|\psi\rangle$. Verify that these two transformations are also unitary.

We can now conclude by summarizing the teleportation protocol: (i) perform a Bell measurement on photons A and B, (ii) send the result of the measurement to the location of photon C, (iii) perform the suitable unitary operation.

4.2 Study of Information Transfer

Now, let us analyze how the unitary transformations affect the results of Bob's measurements on photon C. Suppose Bob measures photon C in the basis defined by

$$\begin{aligned} |+\gamma\rangle &= \cos\gamma|H\rangle + \sin\gamma|V\rangle\,, \\ |-\gamma\rangle &= \cos\gamma|V\rangle - \sin\gamma|H\rangle\,, \end{aligned} \tag{4.3}$$

where γ is an arbitrary angle. Let $P(+\gamma)$ and $P(-\gamma)$ be the probabilities of measuring $|+\gamma\rangle$ and $|-\gamma\rangle$ respectively. If Bob receives the two bits of information from Alice, the statistics are those expected for $|\psi\rangle$, but if he does not, his results look completely random.

Exercise 4.4.

(1) First, we analyze the case where photon A is initially prepared in the state $|\psi\rangle = |H\rangle$. After the protocol as described in the previous section, find $P(+\gamma)$ and $P(-\gamma)$.

(2) Let us now suppose that the result of the Bell-state measurement on A-B was not sent to the location of photon C, hence no unitary transformation was performed before measuring C. What are $P(+\gamma)$ and $P(-\gamma)$?

(3) Repeat points 1 and 2 for the case in which photon A was prepared in the state $|\psi\rangle = |V\rangle$.

(4) A friend of yours, having heard some popular explanations of quantum teleportation, thinks that it can be used to transmit information faster than light. *"It's simple! Because of the Bell-state measurement, the state of photon C changes instantaneously. It's therefore enough for Alice to prepare photon A either in state $|H\rangle$, or in state $|V\rangle$; Bob receives this state instantaneously and can distinguish $|H\rangle$ from $|V\rangle$.* Why is this not possible? Can there be any other protocol that exploits quantum teleportation to send information faster than light?

An interesting point to note about the teleportation protocol described in this chapter is that, information about the original state of photon A did not physically travel between the locations of photons A and C. Instead, it is the measurement on A and B that modifies the state of photon C through entanglement. Alice does not even need to know the polarization of photon A to teleport it over to Bob.

Another important point is that this protocol only allows the teleportation of *information* about a state, and not *matter* itself. It is in principle possible to teleport any quantum state given the presence of enough entangled matter, however in practice this has only been done for photons.

4.3 Summary

Quantum physics allows the teleportation of information without any direct interaction or transfer of mass. In the teleportation protocol, Alice and Bob are respectively given photons B and C belonging to an entangled Bell state. Alice teleports the state of an arbitrary photon A to Bob's photon C by first performing a Bell measurement of photons A and B. She then sends the measurement result to Bob, who performs a unitary transformation on photon C to complete the teleportation. Because of this last step, teleportation cannot be used to send information faster than light.

4.4 The Broader View

4.4.1 *Entanglement swapping*

The question we want to address here is: what happens if the photon to be teleported also belongs to an entangled pair? The situation is sketched in Figure 4.2: photon A starts out being entangled with photon D. The explicit

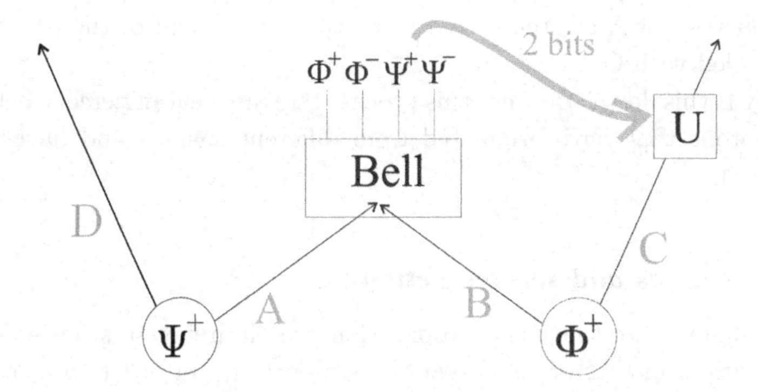

Fig. 4.2 Scheme for entanglement swapping: teleportation of a photon that is itself entangled with another one.

calculation is proposed as an exercise below; the result is at the same time obvious and deep: at the end of the teleportation process, *photon D is entangled with photon C.*

Exercise 4.5. Suppose that photons D-A, B-C are entangled as follows:

$$|\Psi^+\rangle_{DA} = \frac{1}{\sqrt{2}}(|H\rangle|V\rangle + |V\rangle|H\rangle),$$
$$|\Phi^+\rangle_{BC} = \frac{1}{\sqrt{2}}(|H\rangle|H\rangle + |V\rangle|V\rangle).$$

Verify the following equality:

$$|\Psi^+\rangle_{DA}|\Phi^+\rangle_{BC} = \frac{1}{2}|\Phi^+\rangle_{AB}|\Psi^+\rangle_{CD} + \frac{1}{2}|\Phi^-\rangle_{AB}|\Psi^-\rangle_{CD} + \frac{1}{2}|\Psi^+\rangle_{AB}|\Phi^+\rangle_{CD} - \frac{1}{2}|\Psi^-\rangle_{AB}|\Phi^-\rangle_{CD}.$$

If one looks carefully at this process, the situation is: at the beginning, D-A are entangled and B-C are entangled. The Bell measurement creates A-B entanglement, and the result is C-D entanglement. Because of this, the process is called *entanglement swapping.*

Why is this simple? Because the result cannot be anything else! Remember that the state describes the properties of the physical system. Now, the properties of photon A were entangled with those of photon D: if all the properties of A are teleported to C, then at the end of the process, D is entangled with C.

Why is this deep? Because this process generates entanglement between two photons that have originated from different sources and have never interacted!

4.4.2 *Physics and science fiction*

In this chapter, we touched on topics that are often met in science fiction: teleportation and faster-than-light communication. In order to avoid any misunderstanding, let us be explicit about these issues.

We have already said that, in quantum teleportation, it is the *state* that is teleported: in other words, the information about the physical properties of the system. Neither matter nor energy is teleported: a photon must already be present, on which the information can be encoded. In all physics without exception, matter and energy cannot be teleported: they can only be transported, passing from one location to another through all the intermediate locations.

Moreover, this transfer of energy or matter must happen with a speed that does not exceed that of light in vacuum. This limitation also holds for the transfer of *usable* information, as highlighted by an exercise in this chapter: in the teleportation scheme, we do not know what happens (if anything) at the level of the states, but we know that Bob won't see any difference in any of his measurements until he receives the two bits of classical communication. This fact goes beyond the specific teleportation protocol: quantum entanglement can never be used to send a message, neither faster nor even slower than light. We say that *entanglement is a no-signaling resource*. If you want to send a message, you have to send some matter or energy carrying the information.

May these limitations (no teleportation of matter, no signaling faster than light) be overcome one day? In full honesty, we have to say: they have not been *directly* proven to be impossible, in the sense that one may discover unexpected new physical phenomena. If such phenomena do exist, a complete reshaping of physics will be required. In other words, as of today, teleportation of matter and signaling faster than light belong to science fiction and only there: there is absolutely no hint of them in the known degrees of freedom of the universe.

4.5 References and Further Reading

Easy reading:

- A. Zeilinger, Scientific American **282**, 50 (2000).

Resources:

- Original article for teleportation: C.H. Bennett, G. Brassard, C. Crépeau, R.Jozsa, A. Peres, W.K. Wootters, Phys. Rev. Lett. **70**, 1895 (1993).
- Original article for entanglement swapping: M. Żukowski, A. Zeilinger, M.A. Horne, A.K. Ekert, Phys. Rev. Lett. **71**, 4287 (1993).

Suggestions for projects:

- D. Bouwmeester et al., Nature **390**, 575 (1997).
- Y.-H. Kim, S. Kulik, Y. Shih, Phys. Rev. Lett. **86**, 1370 (2001).

4.6 Solutions to the Exercises

Solution 4.1. We can expand the expression on the left to obtain

$$(c|H\rangle + s|V\rangle)\,|\Phi^+\rangle = \frac{1}{\sqrt{2}}\left(c|HHH\rangle + s|VHH\rangle + c|HVV\rangle + s|VVV\rangle\right),$$

where c, s represent $\cos\theta$ and $\sin\theta$ respectively.

Next, using the definitions of the Bell bases in Equation (4.2), similarly expand the expression on the right. Simplifying it, you should obtain the same expression as above.

Solution 4.2. Using the expression on the right side of Equation (4.1), we can fill up the table as follows:

Outcome of measurement A-B	Resulting state of C	Probability
$\lvert\Phi^+\rangle$	$\cos\theta\lvert H\rangle + \sin\theta\lvert V\rangle$	1/4
$\lvert\Phi^-\rangle$	$\cos\theta\lvert H\rangle - \sin\theta\lvert V\rangle$	1/4
$\lvert\Psi^+\rangle$	$\cos\theta\lvert V\rangle + \sin\theta\lvert H\rangle$	1/4
$\lvert\Psi^-\rangle$	$\cos\theta\lvert V\rangle - \sin\theta\lvert H\rangle$	1/4

Solution 4.3.

(1) To verify the transformation is unitary, we calculate the scalar product of the states before and after the transformation and check that they are the same.

Before transformation	After transformation
$\langle H\vert H\rangle = 1$	$\langle T(H)\vert T(H)\rangle = \langle H\vert H\rangle = 1$
$\langle V\vert V\rangle = 1$	$\langle T(V)\vert T(V)\rangle = \langle V\vert V\rangle = 1$
$\langle H\vert V\rangle = 0$	$\langle T(H)\vert T(V)\rangle = \langle H\vert V\rangle = 0$

(2) For $\cos\theta|V\rangle + \sin\theta|H\rangle$, the transformation is $|V\rangle \to |T(V)\rangle = |H\rangle$, $|H\rangle \to |T(H)\rangle = |V\rangle$. This transformation is unitary because

$$\langle T(V)|T(V)\rangle = \langle H|H\rangle = 1\,,$$
$$\langle T(H)|T(H)\rangle = \langle V|V\rangle = 1\,,$$
$$\langle T(H)|T(V)\rangle = \langle V|H\rangle = 0\,.$$

For $\cos\theta|V\rangle - \sin\theta|H\rangle$, the transformation is $|V\rangle \to |T(V)\rangle = |H\rangle$, $|H\rangle \to |T(H)\rangle = -|V\rangle$. This transformation is unitary because

$$\langle T(V)|T(V)\rangle = \langle H|H\rangle = 1\,,$$
$$\langle T(H)|T(H)\rangle = \langle V|V\rangle = 1\,,$$
$$\langle T(H)|T(V)\rangle = -\langle V|H\rangle = 0\,.$$

Hence, we see that these two transformations are possible. This means that with the information on the outcome of Alice's measurement on photons A-B, Bob can transform photon C to obtain the original state of photon A.

Solution 4.4.

(1) At the end of the protocol, the state of photon C is the same as the initial state of photon A, which is $|H\rangle$. Hence

$$P(+\gamma) = |\langle H| + \gamma\rangle|^2 = \cos^2\gamma\,,$$
$$P(-\gamma) = |\langle H| - \gamma\rangle|^2 = \sin^2\gamma\,.$$

(2) Since photon A is prepared in the state $|\psi\rangle = |H\rangle$,

$$|H\rangle_A|\Phi^+\rangle_{BC} = \frac{1}{2}|\Phi^+\rangle_{AB}|H\rangle_C + \frac{1}{2}|\Phi^-\rangle_{AB}|H\rangle_C + \frac{1}{2}|\Psi^+\rangle_{AB}|V\rangle_C + \frac{1}{2}|\Psi^-\rangle_{AB}|V\rangle_C\,.$$

Hence photon C has a $\frac{1}{2}$ probability of being $|H\rangle$ or $|V\rangle$, thus

$$P(+\gamma) = \frac{1}{2}\cos^2\gamma + \frac{1}{2}\sin^2\gamma = \frac{1}{2}\,,$$
$$P(-\gamma) = \frac{1}{2}\cos^2\gamma + \frac{1}{2}\sin^2\gamma = \frac{1}{2}\,.$$

(3) • At the end of the protocol, the state of photon C is $|V\rangle$. Hence

$$P(+\gamma) = |\langle V| + \gamma\rangle|^2 = \sin^2\gamma\,,$$
$$P(-\gamma) = |\langle V| - \gamma\rangle|^2 = \cos^2\gamma\,.$$

• Without applying the unitary transformation,

$$|V\rangle_A|\Phi^+\rangle_{BC} = \frac{1}{2}|\Phi^+\rangle_{AB}|V\rangle_C - \frac{1}{2}|\Phi^-\rangle_{AB}|V\rangle_C + \frac{1}{2}|\Psi^+\rangle_{AB}|H\rangle_C - \frac{1}{2}|\Psi^-\rangle_{AB}|H\rangle_C\,.$$

Hence, similar to question 2,

$$P(+\gamma) = \frac{1}{2},$$
$$P(-\gamma) = \frac{1}{2}.$$

(4) This is not possible because if the unitary transformations are not performed, $P(+\gamma)$ and $P(-\gamma)$ are the same for $|H\rangle$ and $|V\rangle$, as can be seen from the calculations above. Thus Bob would not know if Alice had prepared the photons in $|H\rangle$ or $|V\rangle$. To perform the transformation, information about the Bell-state measurement on A-B has to be sent (through a slower-than-light signal) by Alice to Bob so that the appropriate unitary transformation can be done. Hence quantum teleportation cannot be exploited to send information faster than light.

Solution 4.5. First, expand the expression on the left to obtain

$$|\Psi^+\rangle_{AB}|\Phi^+\rangle_{CD} = \frac{1}{2}\left(|HHHV\rangle + |VHHH\rangle + |HVVV\rangle + |VVVH\rangle\right).$$

Next, using the definitions of the Bell bases in Equation (4.2), similarly expand and simplify the expression on the right side to verify the equality.

ENTANGLED SOURCE B
QUANTUM
100%
HELLO--
CLASSICAL
TELEPORT

SEND CLASSICAL INFORMATION ?
YES
CANCEL
OUCH MY STOMACH...
WC

WE SWEAR IT! WE DON'T HAVE ANY!!
YOU VIOLATED THE BELL INEQUALITIES. SHOW ME YOUR COMMUNICATION DEVICES!
NEWTON

Chapter 5

Quantum Correlations and Bell's Inequality

Entanglement is one of the most important features of quantum physics. In chapter 1, we have introduced this notion formally and mentioned how it defies our everyday intuition. Here we present what is probably the most direct manifestation of entanglement: the correlations between outcomes of measurements on separated physical systems.

Such a possibility was first noticed by Einstein, Podolsky and Rosen (EPR) in 1935; they found it absurd and came up with an argument to try and show that quantum theory was not the whole story and needed to be completed. While EPR indeed put the finger on something extremely interesting, their article and the subsequent reply by Bohr somehow missed the point, which was clarified only in 1964 by Bell. Here, we study directly the approach initiated by Bell; a commentary on the EPR paper will be given in chapter 9.

5.1 Quantum Description

We describe an experimental setup that demonstrates quantum correlations between two entangled photons.

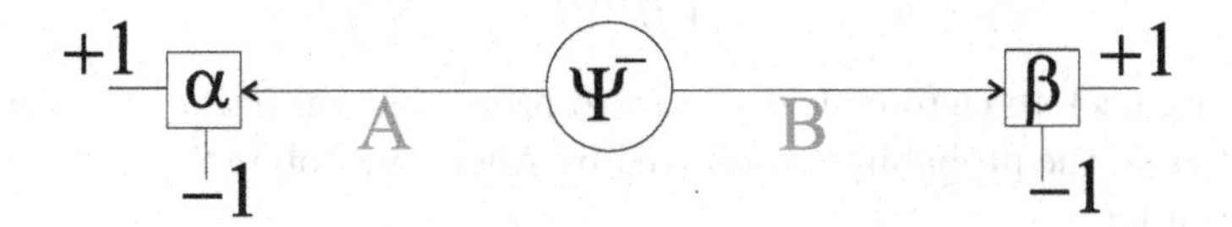

Fig. 5.1 Setup to demonstrate quantum correlations between two entangled photons.

5.1.1 *Experimental setup*

First, a source emits two photons. Alice and Bob each receive one of these photons, labeled as photons A and B respectively. They measure the polarization of their respective photons along the directions α and β (Figure 5.1). Mathematically, this is described by the bases

$$|+\alpha\rangle = \cos\alpha|H\rangle + \sin\alpha|V\rangle\,,$$
$$|-\alpha\rangle = \cos\alpha|V\rangle - \sin\alpha|H\rangle\,,$$

for Alice and

$$|+\beta\rangle = \cos\beta|H\rangle + \sin\beta|V\rangle\,,$$
$$|-\beta\rangle = \cos\beta|V\rangle - \sin\beta|H\rangle\,,$$

for Bob. They can only make a single measurement of each photon, because the process changes the state of the photon such that any further measurements do not give information about the photon's original state.

There are four possible measurement outcomes for each pair of emitted photons. We use $r_A(\alpha)$ and $r_B(\beta)$ to denote the measurement outcomes observed by Alice and Bob respectively. If Alice or Bob measure $|+\alpha\rangle$ or $|+\beta\rangle$, they will record it as $r_A(\alpha) = +1$ or $r_B(\beta) = +1$; if they measure $|-\alpha\rangle$ or $|-\beta\rangle$, they will record it as -1. Then the four possible measurement outcomes will be labeled $(+,+)$, $(+,-)$, $(-,+)$ and $(-,-)$.

This experiment can be carried out for many rounds, with Alice and Bob recording their outcomes. After the experiment, they then find the number of times that each of the four measurement outcomes occur, in order to infer their probabilities. For example, the fraction of measurements in which $r_A(\alpha) = +1$ and $r_B(\beta) = -1$ will translate into a probability $P(+-|\alpha,\beta)$, where the $+$ sign refers to α and the $-$ to β. Similarly, they can infer $P(++|\alpha,\beta)$, $P(-+|\alpha,\beta)$, and $P(--|\alpha,\beta)$.

5.1.2 *Source of entangled photons*

We consider now that the source prepares the *singlet state* defined by

$$|\Psi^-\rangle = \frac{1}{\sqrt{2}}\left(|H\rangle|V\rangle - |V\rangle|H\rangle\right). \tag{5.1}$$

Referring back to chapter 1, it is easy to prove that this state is entangled. In this case, the probabilities observed by Alice and Bob in this experiment are given by

$$\begin{aligned} P(++|\alpha,\beta) &= P(--|\alpha,\beta) = \tfrac{1}{4}\big[1 - \cos 2(\alpha-\beta)\big]\,, \\ P(+-|\alpha,\beta) &= P(-+|\alpha,\beta) = \tfrac{1}{4}\big[1 + \cos 2(\alpha-\beta)\big]\,. \end{aligned} \tag{5.2}$$

Exercise 5.1.
(1) Verify Equation (5.2).
(2) Verify in particular that if $\alpha = \beta$, then $r_A(\alpha) = -r_B(\beta)$ deterministically.
(3) Verify that the average value of $r_A(\alpha)$ is 0 for all α, and similarly for $r_B(\beta)$.

This means that when Alice or Bob take down their measurements individually, they will observe that the two outcomes are equally likely to be measured. In particular, what Alice or Bob observes locally is not affected by what the other person does. This result is not surprising; we have already seen in chapter 4.2 that entangled photons cannot be used to transmit signals.

However, the interesting thing is that when they share their results, they will find that the probability of each outcome is dependent on *both* their measurement bases. The photon reaching Alice produces its outcome depending on what measurement the other photon at Bob undergoes!

Note that we have made no mention of the order in which Alice and Bob make their measurements. Neither have we specified their distance apart or the time interval between their measurements. These factors simply do not affect the quantum correlations between the entangled photons: even if they were measured at the same time, thousands of miles apart, they would still show this correlated behavior!

Now, we introduce a measure of the degree by which the photons are correlated: the *correlation coefficient*, defined as

$$E(\alpha, \beta) = P(r_A(\alpha) = r_B(\beta)) - P(r_A(\alpha) \neq r_B(\beta)) . \tag{5.3}$$

This can have a range of values, from -1 to $+1$. A value of -1, also known as *anti-correlation*, means that when one photon gives an outcome, the other would always give the opposite outcome. On the other hand, a value of $+1$, or perfect *correlation*, means that the two photons will always give the same outcome.

For the probabilities in Equation (5.2), the correlation coefficient is given by

$$E(\alpha, \beta) = -\cos[2(\alpha - \beta)] . \tag{5.4}$$

Exercise 5.2.

(1) Verify Equation (5.4).

(2) Verify that when the outcomes are labeled as +1 and −1 as in this experiment, the correlation coefficient can also be written as the average value of the product $r_A(\alpha)\, r_B(\beta)$:

$$E(\alpha, \beta) = \langle r_A(\alpha)\, r_B(\beta) \rangle \,. \tag{5.5}$$

5.1.3 *Mechanism for correlations?*

We have seen that each photon produces an outcome that is correlated with that produced by the other photon, even if they are very far apart. How is this possible, is there a *mechanism* that can explain this strange correlation at a distance? Did they communicate with each other, or did they make some agreement at the source?

While quantum physics is able to accurately predict the probabilities given above for each measurement outcome, it does not propose any mechanism to explain these correlations. Einstein and many others perceived this feature as a failure of the theory. It is actually one of its greatest successes; experiments vindicated the fact that there is actually no such mechanism, as we will present in this chapter. In order to reach this remarkable conclusion, we have to study the possible candidates for a mechanism and rule them out.

5.2 Attempts at Classical Explanations

There are only two possible classical mechanisms that explain correlations between distant events:

(1) **Correlation by communication**
The first possible mechanism is that the photons may *communicate* with each other by exchanging signals. For example, photon A, which is measured first, may send some information about the measurement α and its outcome $r_A(\alpha)$ to photon B, which would then produce its outcome accordingly.

(2) **Correlation by pre-established agreement**
The second possible mechanism is *pre-established agreement*: since both photons were produced by the same source, they could be carrying

some common information about what they would each do for any measurement they encountered. In other words, we assume that each of them carries a list

$$\lambda_A = \{..., r_A(\alpha), r_A(\alpha'), r_A(\alpha''), ...\},$$
$$\lambda_B = \{..., r_B(\beta), r_B(\beta'), r_B(\beta''), ...\}.$$

These lists specify the outcomes for each possible measurement, and are also commonly known as *hidden variables.* For example, if λ_A and λ_B are such that $r_A(x) = -r_B(x)$ for all x, then when $\alpha = \beta$, the photons will always give opposite results when measured in the same direction. Thus hidden variables can reproduce the perfect anti-correlations noticed in the previous section.

The first of these mechanisms can be experimentally ruled out; we have already given the clues for this and we leave it as an exercise. The next section will describe how the second mechanism can be ruled out.

Exercise 5.3. Propose an experimental arrangement that can rule out correlation by communication.

5.3 Bell's Theorem

Is there any way to determine whether photons correlate at the source using hidden variables and pre-established agreement? This problem puzzled physicists for almost 30 years, until an ingenious test was proposed by John S. Bell in 1964. Many variants of Bell's theorem have since been formulated; here we will discuss the CHSH inequality, derived by Clauser, Horne, Shimony and Holt in 1969.

We start by modifying the experimental procedure described earlier. Previously, when the source emits two photons, Alice and Bob each use one measurement basis. Now, they choose between two different measurement bases: α and α' for Alice, β and β' for Bob.

Suppose that the photons exchange an agreement at the source about the outcome that they would each produce. This is equivalent to them forming the hidden variable $\lambda = \{\lambda_A, \lambda_B\}$, where

$$\lambda_A = \{r_A(\alpha), r_A(\alpha')\},$$
$$\lambda_B = \{r_B(\beta), r_B(\beta')\}.$$

Having these numbers, we can form the expression

$$S(\lambda) = \left(r_A(\alpha) + r_A(\alpha')\right) r_B(\beta) + \left(r_A(\alpha) - r_A(\alpha')\right) r_B(\beta') . \quad (5.6)$$

It can be verified that for every possible λ, $S(\lambda)$ can only take the values $+2$ or -2.

Exercise 5.4. Verify that $S(\lambda)$ can only take the values $+2$ or -2.

In a single run of the experiment, $S(\lambda)$ cannot be measured. Indeed, we note that Alice would have to measure photon A using both the α and α' bases; similarly for Bob. This is impossible because they can only perform one measurement on each photon. Nevertheless, some information about $S(\lambda)$ can be obtained: if they repeat the experiment with many pairs of photons, the *average value* of S can be determined:

$$\langle S \rangle = E(\alpha, \beta) + E(\alpha', \beta) + E(\alpha, \beta') - E(\alpha', \beta') . \quad (5.7)$$

Exercise 5.5. Verify Equation (5.7) by noticing that

$$\begin{aligned} \langle S \rangle = & \langle r_A(\alpha)\, r_B(\beta) \rangle + \langle r_A(\alpha')\, r_B(\beta) \rangle \\ & + \langle r_A(\alpha)\, r_B(\beta') \rangle - \langle r_A(\alpha')\, r_B(\beta') \rangle . \end{aligned} \quad (5.8)$$

Hint: use the definition of the average $\langle x \rangle = \lim_{N\to\infty} \frac{1}{N} \sum_{k=1}^{N} x_k$.

Using the current experimental setup, Alice and Bob can obtain all four correlation coefficients in Equation (5.7), and hence estimate $\langle S \rangle$.

If the assumption of hidden variables is correct, then $\langle S \rangle$ is the average value of a number that can take only the values $+2$ or -2. So obviously,

$$|\langle S \rangle| \leq 2 . \quad (5.9)$$

This relation is called *Bell's inequality.*

We thus find a way to verify if the photons indeed use hidden variables as a mechanism for their correlated behavior: if Bell's inequality is satisfied, then it can be a possible mechanism. If not, we would have to conclude that pre-established agreement at the source does not explain quantum correlations.

When one inserts the parameters for the above experiment, one can indeed verify that Bell's inequality can be violated for certain choices of measurement bases, as in the following exercise.

Exercise 5.6.

(1) Using the quantum expression (5.4) of $E(\alpha, \beta)$, find an expression for (5.7) in this experiment. Verify that Bell's inequality (5.9) is *violated* for suitable choices of measurement bases. *Hint:* try $\alpha = 0$, $\alpha' = \frac{\pi}{4}$, $\beta = \frac{\pi}{8}$ and $\beta' = -\frac{\pi}{8}$.

(2) Find other measurements directions for which Bell's inequality is violated; find also some measurement directions for which no violation is observed, i.e. $|\langle S \rangle| \leq 2$.

In conclusion, the correlations between entangled photons are not derived from pre-established agreements at the source: hidden variables do not exist. Entangled photons can establish instantaneous correlations with each other even if they are far apart, without any form of communication.

5.4 Summary

Entangled states exhibit quantum correlations that cannot be explained by classical mechanisms of communication or hidden variables. The latter has been tested using Bell's theorem, in which entangled states are able to violate Bell's inequality, thus proving that hidden variables do not exist.

5.5 The Broader View

5.5.1 *The danger of words*

Facing an unexpected phenomenon as the violation of Bell's inequality, it is normal that physicists had a hard time coming up with suitable expressions to describe it. In the main text, we have been as accurate as one could. Other expressions are less accurate, but are widely used in literature, so we mention them here for the sake of those who would like to read more.

Historically, $\lambda = \{\lambda_A, \lambda_B\}$ was called a *local hidden variable*: "local" because it would allow outcomes to be determined using only information that is available at the measurement location; "hidden" because it does not appear in quantum theory. The "hidden" part is rather imprecise: nowhere, in the derivation of Bell's inequality, does one make the assumption that the pre-established agreement must be hidden! In other words, the violation of

Bell's inequality falsifies any kind of pre-established agreement, be it hidden or public.

The fact that Bell's inequality is violated is often summarized in two expressions. Both are harmless if one knows what is being spoken about, but may convey wrong (or highly controversial) ideas to the beginner.

- *Violation of local realism.* If "local realism" is a shortening of the expression "the assumption of the pre-existence of measurement results", the term is perfectly adequate. An equivocation may arise with the philosophical notion of realism: realism in philosophy indicates the standpoint in which one believes in the existence of an external world and in the possibility of knowing it well through experience; it is a sort of mid-point between empiricism[1] and rationalism[2]. Now, it would be a bit far-fetched to conclude "non-realism" from the violation of Bell's inequality, in the sense that "there is no external reality": on the contrary, if there is no external reality, any experience (including the violation of Bell's inequality) loses its cogency for knowledge. What is certainly true is: a philosophical realist, after learning about the violation of Bell's inequality, cannot be a "local realist", i.e. cannot believe in pre-established agreement for results of quantum measurements.
- *Non-locality.* Here as well, if by "non-locality" one means that quantum correlations cannot be due to pre-established agreements followed by readouts that only take into account locally available information, the term is perfectly adequate. However, the term may convey the idea of an actual "effect at a distance" of one measurement on the other. There is certainly no experimental evidence for such an effect, nor any need for it in the theory.

The unexperienced person may be astonished at how strongly some physicists may feel in favor of or against any of these two expressions. If one day, speaking about these topics, you use the "wrong" word and someone attacks you on that, just keep your calm, define your terms or change your words: everything will come back to normal.

---

[1]In empiricism, our knowledge is just the sum of experiences, without any higher order or structure.

[2]In rationalism, experiences are worthless and only mathematical thought leads to rigorous knowledge.

5.5.2 *Incompatible physical quantities*

The violation of Bell's inequality casts a clear light on a very general notion in quantum physics: that of *incompatible measurements*, or more precisely *incompatible physical quantities*. This notion dates back to the early years of quantum mechanics: it underlies the famous Heisenberg's uncertainty relations for position and momentum, to which we shall return briefly in chapter 9.

When we say that two physical quantities A and A' are incompatible, we mean the following: suppose that a is a possible outcome for the measurement of A, and denote the corresponding state by $|a\rangle$. On such a state, in general, the value of A' is not well-defined: if A' is measured on the state $|a\rangle$, several results are possible and the outcome is probabilistic.

Let us take an example with a photon. Let the first physical quantity A be the polarization in the $\{|H\rangle, |V\rangle\}$ basis, and let the second physical quantity A' be the polarization in another basis, say for instance $\{|+\rangle$, $|-\rangle\}$ where $|\pm\rangle \equiv |\alpha = \pm\frac{\pi}{4}\rangle$. Take now a state associated to an outcome of A: for instance, the state $|H\rangle$. We know that this state is not also associated to an outcome of A': actually, in the example, both outcomes of A' can happen with equal probability. Another way of viewing this is: there is no state $|\text{"}H \text{ and } +\text{"}\rangle$ that would give both outcome H for A and $+$ for A' with certainty. Thus these two physical quantities are incompatible.

In summary, not all physical quantities can be assigned a definite value simultaneously. For polarization, only *one* basis can be determined. When one studies more complex degrees of freedom, it is possible that a few physical quantities can be simultaneously determined (for instance, energy and angular momentum), but not all of them. For instance, position and momentum can never be simultaneously determined.

This is basic quantum physics, but how do we know that it is really the case? One could argue that maybe we have not found the right measurement procedure, or the right theory, and we are just erecting our ignorance or lack of fantasy to the status of physical law... There are many answers to this objection, but the most direct one is provided precisely by the violation of Bell's inequality. If all measurements were compatible and we just did not notice it, it would be impossible to violate Bell's inequality, because all the possible measurement outcomes would exist. But the inequality *is* violated in experiments. Therefore there are incompatible measurements.

A final remark: in discussions on the topic of incompatible measurements, a different result is often quoted as well: the Kochen-Specker theorem. We cannot discuss it here, since its presentation requires a slightly higher level of formalism than the one introduced in chapter 1.

5.5.3 *True randomness, true secrecy*

We have stressed strongly enough what the violation of Bell's inequality means: there is no pre-established agreement; the photons do not leave the source with a pre-arranged list of outcomes for all possible measurements. Now, this means that *the outcomes of such measurements are true random numbers*. Why? Read these sentences again, it's almost obvious. Indeed, "random" means "not determined": it means that the numbers are generated by an unpredictable process and are not the results of a given algorithm. Now, if there was an algorithm generating those numbers, the algorithm could have produced a list before the experiment started: this is precisely what Bell's inequality check for! So, *the violation of Bell's inequality guarantees the generation of true random numbers*.

The idea can be given yet another twist, namely: *these random numbers are also secret*; nobody else can have the same list. The argument is again the same: if a third party could have a list, we could put this third party very close to the source, produce the list and encode it in the photons before they propagate far away from one another. But then again, the photons would be acting according to a pre-existing list, which is impossible if Bell's inequality is violated.

Note now that Bell's inequality is independent of quantum physics: if it is violated, there is no list, period. Therefore, we can check for true randomness or true secrecy of a *black-box*: we do not need to know anything about how it works, if it uses photons or neutrons or anything else, if it measures any specific polarization... In this sense, Bell's inequality is not only a test of the foundations of quantum physics: it can be given a practical application in random number generation and cryptography. We cannot explore this direction further here, but some references for personal study are given on the next page.

5.5.4 *About loopholes*

Those who read further may at some point encounter some mention about *loopholes* in the experiments that showed a violation of Bell's inequality.

These loopholes have their importance, but not in the sense that they may undermine all that we have discussed. We devote some short comments to them, in order for the reader to acquire the right perspective.

The main loopholes are the following:

- *Locality loophole.* This happens when the measurements of Alice and Bob are not chosen randomly in each run of the experiment, or more precisely, when information about the measurement chosen by Alice could have reached Bob's lab at the speed of light, before Bob's particle is measured (or vice versa). In this situation, one might attribute the correlations to an unknown signal propagating at the speed of light.
- *Detection loophole.* Detectors have limited efficiency: sometimes, even if a photon arrives, they may not fire. This loophole assumes that the detector's firing or not may depend on the measurement that is being performed, i.e. on the choice of the polarization to be measured. If the efficiency of the detectors is not well above 50%, with such a mechanism one can fake the violation of Bell's inequality with suitable pre-established strategies.

There is no point in giving further details here, about how to close each loophole. The locality loophole has been closed in several experiments with photons. The detection loophole has been closed as well, but only with atoms: these being very close to one another, the same experiment could not close the locality loophole. Thus, at the moment of writing, a loophole free experiment is still missing. May this undermine all that we have written? Certainly not! For experimental physicists, a detector is not a black box: they know exactly how it works, and the reasons for its inefficiency. Nobody thinks that the predictions of quantum theory about the violation of Bell's inequality will be shattered as soon as we have better detector!

However, it is important to close the detection loophole in the black-box assessment mentioned above: if you only have a black-box, then you don't know how your detectors work, so you must close all the loopholes to be sure. The fact that this loophole is hard to close is the reason why quantum-certified black-boxes are not yet available on the market.

5.6 References and Further Reading

Easy reading:

- N.D. Mermin, Am. J. Phys. **49**, 940 (1981).
- A.K. Ekert, Physics World, September 2009, pp 28-32.

Resources:

- Technical and less technical articles by John Bell: J.S. Bell, *Speakable and unspeakable in quantum mechanics* (Cambridge University Press, Cambridge, 1987; 2nd edn 2004).
- Review article on experiments: W. Tittel, G. Weihs, Quantum Inf. Comput. **1**, 3 (2001).
- Lecture notes with several references: V. Scarani, arXiv:0910.4222 (lectures 4-6).

Suggestions for projects:

- G. Weihs et al., Phys. Rev. Lett. **81**, 5039 (1998).
- W. Tittel et al., Phys. Rev. Lett. **81**, 3563 (1998). Note: this experiment does not use polarization as the entangled degree of freedom.

5.7 Solutions to the Exercises

Solution 5.1.

(1) $P(++|\alpha,\beta)$ is the probability of finding the original singlet state $|\Psi^-\rangle$ in the state $|+\alpha\rangle|+\beta\rangle$. Thus we first find an expression for the latter:

$$\begin{aligned}|+\alpha\rangle\otimes|+\beta\rangle &= (\cos\alpha|H\rangle+\sin\alpha|V\rangle)\otimes(\cos\beta|H\rangle+\sin\beta|V\rangle)\\ &= \cos\alpha\cos\beta\,|HH\rangle+\cos\alpha\sin\beta\,|HV\rangle\\ &\quad+\sin\alpha\cos\beta\,|VH\rangle+\sin\alpha\sin\beta\,|VV\rangle\,.\end{aligned}$$

The probability can then be calculated using Born's rule for probabilities:

$$\begin{aligned}P(++|\alpha,\beta) &= \left|\langle+\alpha,+\beta|\Psi^-\rangle\right|^2 = \frac{1}{2}\sin^2(\alpha-\beta)\\ &= \frac{1}{4}\left[1-\cos 2(\alpha-\beta)\right].\end{aligned}$$

Repeating this procedure, we obtain

$$P(++|\alpha,\beta) = P(--|\alpha,\beta) = \frac{1}{4}[1-\cos 2(\alpha-\beta)]\,,$$
$$P(+-|\alpha,\beta) = P(-+|\alpha,\beta) = \frac{1}{4}[1+\cos 2(\alpha-\beta)]\,.$$

(2) If $\alpha=\beta$, then $\alpha-\beta=0$. We can substitute this into the expressions for the probabilities obtained in part (1):

$$P(++|\alpha,\beta) = P(--|\alpha,\beta) = 0\,,$$
$$P(+-|\alpha,\beta) = P(-+|\alpha,\beta) = \frac{1}{2}\,.$$

Thus the results for photons A and B would always be opposite, i.e. $r_A(\alpha) = -r_B(\beta)$.

(3) From the results of part (1), we can calculate

$$\begin{aligned}\langle r_A(\alpha)\rangle &= (+1)\big[P(++|\alpha,\beta) + P(+-|\alpha,\beta)\big]\\ &\quad +(-1)\big[P(--|\alpha,\beta) + P(-+|\alpha,\beta)\big]\\ &= 0\,,\\ \langle r_B(\beta)\rangle &= (+1)\big[P(++|\alpha,\beta) + P(-+|\alpha,\beta)\big]\\ &\quad +(-1)\big[P(--|\alpha,\beta) + P(+-|\alpha,\beta)\big]\\ &= 0\,.\end{aligned}$$

Solution 5.2.

(1) From Equation (5.3), we can substitute the results in Equation (5.2) to obtain:

$$\begin{aligned}E(\alpha,\beta) &= P(++|\alpha,\beta) + P(--|\alpha,\beta) - P(+-|\alpha,\beta) - P(-+|\alpha,\beta)\\ &= \frac{1}{2}[1-\cos 2(\alpha-\beta)] - \frac{1}{2}[1+\cos 2(\alpha-\beta)]\\ &= -\cos[2(\alpha-\beta)]\,.\end{aligned}$$

(2) To prove that $E(\alpha,\beta) = \langle r_A(\alpha)\, r_B(\beta)\rangle$, we note that $r_A(\alpha)\, r_B(\beta)$ is equal to $+1$ when $r_A = r_B$ and -1 when $r_A \neq r_B$. Thus

$$\begin{aligned}\langle r_A(\alpha)\, r_B(\beta)\rangle &= (+1)\big[P(++|\alpha,\beta) + P(--|\alpha,\beta)\big]\\ &\quad +(-1)\big[P(+-|\alpha,\beta) + P(-+|\alpha,\beta)\big]\\ &= E(\alpha,\beta)\,.\end{aligned}$$

Solution 5.3. If Alice and Bob are stationed very far apart, make their measurements at the same time, and choose their measurement bases while the photons are traveling towards them, then there is not enough time for any signal to travel from one photon to the other. Under such an experimental arrangement, if correlations are still observed between the photons, then communication can be ruled out as a mechanism. This has been experimentally verified to a high degree of certainty.

Solution 5.4. We first note that r_A and r_B can only take the values $+1$ or -1. Substituting the possible combinations of values of $r_A(\alpha)$ and $r_A(\alpha')$ into Equation (5.6), we find that if $r_A(\alpha) + r_A(\alpha') = \pm 2$, then $r_A(\alpha) - r_A(\alpha') = 0$ and vice versa. Thus we observe that $S(\lambda)$ can only take the values $+2$ or -2.

Solution 5.5. Using the definition of the average value, we can compute

$$\begin{aligned}\langle S\rangle &= \lim_{N\to\infty}\frac{1}{N}\sum_{k=1}^{N} S_k \\ &= \lim_{N\to\infty}\frac{1}{N}\sum_{k=1}^{N}\big[r_A(\alpha)_k r_B(\beta)_k + r_A(\alpha')_k r_B(\beta)_k \\ &\quad + r_A(\alpha)_k r_B(\beta')_k - r_A(\alpha')_k r_B(\beta')_k\big] \\ &= \langle r_A(\alpha) r_B(\beta)\rangle + \langle r_A(\alpha') r_B(\beta)\rangle + \langle r_A(\alpha) r_B(\beta')\rangle - \langle r_A(\alpha') r_B(\beta')\rangle\,.\end{aligned}$$

From Equation (5.5), we have $E(\alpha,\beta) = \langle r_A(\alpha) r_B(\beta)\rangle$. We can substitute this into the above expression to obtain

$$\begin{aligned}\langle S\rangle &= \langle r_A(\alpha) r_B(\beta)\rangle + \langle r_A(\alpha') r_B(\beta)\rangle + \langle r_A(\alpha) r_B(\beta')\rangle - \langle r_A(\alpha') r_B(\beta')\rangle \\ &= E(\alpha,\beta) + E(\alpha',\beta) + E(\alpha,\beta') - E(\alpha',\beta')\,.\end{aligned}$$

Solution 5.6.

(1) Taking $\alpha = 0$, $\alpha' = \frac{\pi}{4}$, $\beta = \frac{\pi}{8}$ and $\beta' = -\frac{\pi}{8}$, we can substitute these values into Equation (5.4) to obtain:

$$\begin{aligned}E(\alpha,\beta) &= -\cos\left(-\frac{\pi}{4}\right) = -\frac{\sqrt{2}}{2}\,, \\ E(\alpha,\beta') &= -\cos\left(\frac{\pi}{4}\right) = -\frac{\sqrt{2}}{2}\,, \\ E(\alpha',\beta) &= -\cos\left(\frac{\pi}{4}\right) = -\frac{\sqrt{2}}{2}\,, \\ E(\alpha',\beta') &= -\cos\left(\frac{3\pi}{4}\right) = \frac{\sqrt{2}}{2}\,.\end{aligned}$$

From Equation (5.7), we can compute

$$|\langle S\rangle| = 4\times\frac{\sqrt{2}}{2} = 2\sqrt{2} > 2\,.$$

Hence we see that Bell's inequality is violated.

(2) Another set of angles that violate Bell's inequality by giving $\langle S \rangle = 2\sqrt{2}$ is $\alpha = \frac{\pi}{3}$, $\alpha' = \frac{7\pi}{12}$, $\beta = \frac{11\pi}{24}$ and $\beta' = \frac{5\pi}{24}$. On the other hand, $\langle S \rangle = 2$ and no violation is observed when we use the angles $\alpha = 0$, $\alpha' = 2\pi$, $\beta = \pi$ and $\beta' = -\pi$.

OLYMPIAD
a+b+c=even
a+b'+c'=odd
a'+b+c'=odd
a'+b'+c=odd
HMM...THIS LOOKS HARD...HOW AM I SUPPOSED TO START?

Chapter 6

The GHZ Argument for Quantum Correlations

Is violating Bell's inequality the only way to prove that quantum correlations are not due to pre-established agreement? In this chapter, we discuss the Greenberger-Horne-Zeilinger (GHZ) argument, which is a much simpler argument that does not involve inequality. The price to pay is that it involves three entangled particles instead of two.

6.1 Quantum Description

6.1.1 *The setup*

Instead of two photons, we shall now consider a source that produces three

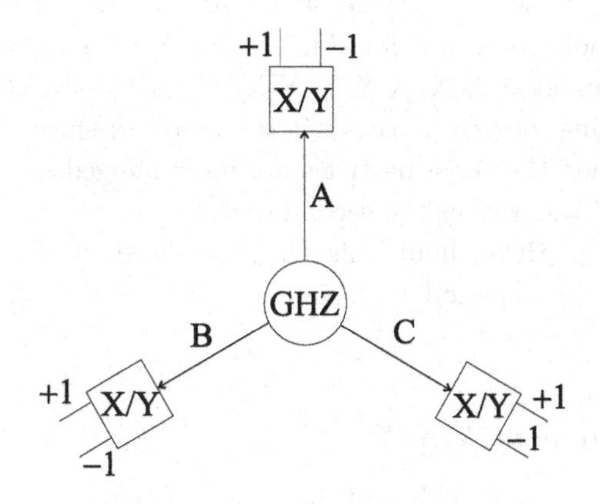

Fig. 6.1 Setup for the GHZ argument.

photons in the entangled state

$$|GHZ\rangle = \frac{1}{\sqrt{2}}\left(|H\rangle|H\rangle|H\rangle + |V\rangle|V\rangle|V\rangle\right). \quad (6.1)$$

The three physicists Alice, Bob and Claire receive photons A, B and C respectively. They can choose between two measurement bases: the "diagonal" X basis or the "circular" Y basis, defined as:

$$\text{X basis:} \quad |\pm x\rangle = \frac{1}{\sqrt{2}}(|H\rangle \pm |V\rangle). \quad (6.2)$$

$$\text{Y basis:} \quad |\pm y\rangle = \frac{1}{\sqrt{2}}(|H\rangle \pm i\,|V\rangle). \quad (6.3)$$

Note that for this calculation, we need a basis defined with a complex number ($i = \sqrt{-1}$).

Similar to the experiment in the previous chapter, we denote the measurement outcomes of Alice, Bob and Claire as r_A, r_B and r_C respectively. If the photon produces the outcome corresponding to the states $|+x\rangle$ or $|+y\rangle$, this will be recorded as $+1$. Otherwise, it will be recorded as -1.

There are eight different ways in which Alice, Bob and Claire can make their choice of polarizers, but for this argument we only need to consider four of them, namely X-X-X, X-Y-Y, Y-X-Y and Y-Y-X. Of these, we will only be considering the first two cases in this chapter: since the state $|GHZ\rangle$ is symmetric when the three particles are interchanged, the last two cases produce similar statistics as the second case.

For simplicity, throughout this chapter we shall use notations like $|+x,-x,+x\rangle \equiv |+x\rangle|-x\rangle|+x\rangle$.

6.1.2 *Measurement X-X-X*

We first study the case in which all three physicists measure in the X basis. It can be verified that the results of their measurements are always arranged such that their product is +1:

$$r_A(x)\, r_B(x)\, r_C(x) = +1\,. \tag{6.4}$$

This shows strong correlations among the three particles. For example, if two of the photons produced the result −1, the last photon will certainly produce the result +1.

Exercise 6.1.

(1) Verify that the statistics of the outcomes are:

$$\begin{aligned}
P(+++|xxx) &= |\langle +x, +x, +x|GHZ\rangle|^2 = \tfrac{1}{4} \\
P(++-|xxx) &= |\langle +x, +x, -x|GHZ\rangle|^2 = 0 \\
P(+-+|xxx) &= |\langle +x, -x, +x|GHZ\rangle|^2 = 0 \\
P(+--|xxx) &= |\langle +x, -x, -x|GHZ\rangle|^2 = \tfrac{1}{4} \\
P(-++|xxx) &= |\langle -x, +x, +x|GHZ\rangle|^2 = 0 \\
P(-+-|xxx) &= |\langle -x, +x, -x|GHZ\rangle|^2 = \tfrac{1}{4} \\
P(--+|xxx) &= |\langle -x, -x, +x|GHZ\rangle|^2 = \tfrac{1}{4} \\
P(---|xxx) &= |\langle -x, -x, -x|GHZ\rangle|^2 = 0\,.
\end{aligned} \tag{6.5}$$

(2) Verify that these probabilities imply Equation (6.4).

Thus we see that three photons produce correlations. Now we ask, are there similar correlations between two of these photons? What are the statistics of one photon? These questions do not have any immediate bearing on the GHZ argument, and are dealt with in the following exercise.

Exercise 6.2.

(1) We first consider what happens if Alice studies her own results before making comparisons with Bob and Claire. Verify that Equation (6.5) implies

$$P(+|x) = P(-|x) = \frac{1}{2}, \tag{6.6}$$

and the same for Bob and Claire. This means that the results measured by each physicist appears to be random, with the photon having an equal probability of producing each outcome.

(2) Now we consider correlations between two photons. Suppose that only Alice and Bob compare their results. Verify the following probabilities:

$$\begin{aligned} P(++\,|xx) &= \tfrac{1}{4}\,, \\ P(+-\,|xx) &= \tfrac{1}{4}\,, \\ P(-+\,|xx) &= \tfrac{1}{4}\,, \\ P(--\,|xx) &= \tfrac{1}{4}\,. \end{aligned} \tag{6.7}$$

By the symmetry of the experimental setup, these probabilities are the same if Alice and Claire or Bob and Claire compare their results.

6.1.3 *Measurement X-Y-Y*

In the previous section, we have observed quantum correlations between three photons when the measurement X-X-X was used. How about the measurement X-Y-Y, in which one of the physicists measures in the X basis, while the others measure in the Y basis?

We can show that correlations are again present between the three photons:

$$r_A(x)\, r_B(y)\, r_C(y) = -1\,, \tag{6.8}$$

$$r_A(y)\, r_B(x)\, r_C(y) = -1\,, \tag{6.9}$$

$$r_A(y)\, r_B(y)\, r_C(x) = -1\,, \tag{6.10}$$

with certainty. This means that when one physicist measures in the X basis and the other two physicists measure in the Y basis, the three results are always arranged such that their product is -1.

Exercise 6.3.

(1) Suppose that Alice measures in the X basis. Verify that the probabilities of all the possible outcomes are

$$\begin{aligned}
P(+++|xyy) &= \left|\langle +x,+y,+y|GHZ\rangle\right|^2 = 0\,,\\
P(++-|xyy) &= \left|\langle +x,+y,-y|GHZ\rangle\right|^2 = \tfrac{1}{4}\,,\\
P(+-+|xyy) &= \left|\langle +x,-y,+y|GHZ\rangle\right|^2 = \tfrac{1}{4}\,,\\
P(+--|xyy) &= \left|\langle +x,-y,-y|GHZ\rangle\right|^2 = 0\,,\\
P(-++|xyy) &= \left|\langle -x,+y,+y|GHZ\rangle\right|^2 = \tfrac{1}{4}\,,\\
P(-+-|xyy) &= \left|\langle -x,+y,-y|GHZ\rangle\right|^2 = 0\,,\\
P(--+|xyy) &= \left|\langle -x,-y,+y|GHZ\rangle\right|^2 = 0\,,\\
P(---|xyy) &= \left|\langle -x,-y,-y|GHZ\rangle\right|^2 = \tfrac{1}{4}\,.
\end{aligned} \tag{6.11}$$

By the symmetry of the GHZ state and the experimental setup, the statistics are the same if the physicist measuring in the X basis is Bob or Claire.

(2) Verify that the statistics above imply Equations (6.8), (6.9) and (6.10).

6.2 Impossibility of Classical Mechanisms for Correlations

Having observed strong correlations for the outcomes of some measurements on the GHZ state, we are now going to consider possible classical mechanisms for the behavior of these photons.

(1) **Correlation by communication**
With reference to the previous chapter, it can be verified that these correlations are not due to communication between the photons.

(2) **Correlation by pre-established agreement**
We now suppose that the photons have "pre-established agreements" at the source about the measurement results that they would produce. Mathematically, this is equivalent to the photons leaving the source with a common list of six numbers

$$\lambda = \{r_A(x), r_A(y);\ r_B(x), r_B(y);\ r_C(x), r_C(y)\}\,. \tag{6.12}$$

It can be verified that no such list exists, which means that pre-established agreement cannot explain the correlations observed among the photons. This argument leads to the same conclusion as that in the previous section using Bell's inequality.

Exercise 6.4. Verify that no list (6.12) can fulfill the four conditions (6.4), (6.8), (6.9) and (6.10).

6.3 Summary

The GHZ-argument, involving an entangled state of three particles, is another proof that quantum correlations are not derived from pre-established agreement.

6.4 The Broader View

The GHZ argument has been presented as a useful exercise to practice the formalism of quantum physics. Its message, however, is ultimately the same as for Bell's inequality: the outcomes of quantum measurements do not come from pre-established agreement. Yet another argument for the same conclusion, with a different flavor, is known as *Hardy's paradox.* Instead of insisting again on the same message, we introduce here a different kind of test, that allows checking for a more subtle aspect of quantum theory.

6.4.1 *Falsification of Leggett's model*

Let us go back to the very definition of entanglement given in chapter 1: a state of two systems that cannot be written as a product of two states, one for each system. We stressed the difference between an entangled state in quantum physics and the state of a composite classical system like the Earth and Moon: within the theory, this means that *individual systems do not have well-defined properties.*

Now, can one test this last statement directly? You may think that Bell's inequality do that, but actually pre-established agreement is a stronger requirement than just the existence of some individual properties. A more refined test was proposed by Leggett and improved in subsequent works. Here we can only give the gist of it.

As usual, two photons are sent apart from one another and measured at different locations, one by Alice and the other by Bob. Leggett's model assumes that there is a hidden parameter λ that plays the following role: if Alice knows λ, she would notice that, for any fixed value of λ, her photon gives the statistics corresponding to a well-defined one-photon state $|\alpha_\lambda\rangle$. Similarly for Bob: for any fixed value of λ, his photon gives the statistics corresponding to a well-defined one-photon state $|\beta_\lambda\rangle$. It is *not* assumed that the correlations should be compatible with the state $|\alpha_\lambda\rangle \otimes |\beta_\lambda\rangle$, because such a model would not violate Bell's inequality. In fact, the model does not assume anything about the correlations: the only aim is to try and save, at least, the existence of some well-defined properties for individual photons.

Even such a general model has been tested and falsified. The criterion, like Bell's inequality, is independent of quantum physics: therefore, we can really say that, in some cases (corresponding to what quantum theory describes as entangled states), individual systems do lose their individual properties.

6.5 References and Further Reading

Easy reading:

- D.M. Greenberger, M. Horne, A. Zeilinger, Physics Today, August 1993, pp 22-29.

Resources:

- D.M. Greenberger, M. Horne, A. Zeilinger, in: E. Kafatos (ed.), *Bell's Theorem, Quantum Theory, and Conceptions of the Universe* (Kluwer, Dordrecht, 1989), p.69.
- N.D. Mermin, Am. J. Phys. **58**, 731 (1990).
- On Leggett's model: S. Gröblacher et al., Nature **446**, 871 (2007); C. Branciard et al., Nature Physics **4**, 681 (2008).

Suggestion for projects:

- J.-W. Pan et al., Nature **403**, 515 (2000).

6.6 Solutions to the Exercises

Solution 6.1.

(1) $P(+++|xxx)$ is the probability that all three photons will be transmitted when Alice, Bob and Claire make measurements using the $|\pm x\rangle$ basis. We compute this by first expanding the expression

$$\begin{aligned} &|+x,+x,+x\rangle \\ &= \left(\frac{1}{\sqrt{2}}(|H\rangle + |V\rangle)\right)\left(\frac{1}{\sqrt{2}}(|H\rangle + |V\rangle)\right)\left(\frac{1}{\sqrt{2}}(|H\rangle + |V\rangle)\right) \\ &= \frac{1}{2\sqrt{2}}(|HHH\rangle + |HHV\rangle + |HVH\rangle + |HVV\rangle \\ &\quad +|VHH\rangle + |VHV\rangle + |VVH\rangle + |VVV\rangle)\,. \end{aligned}$$

Since $|GHZ\rangle = \frac{1}{\sqrt{2}}(|H\rangle|H\rangle|H\rangle + |V\rangle|V\rangle|V\rangle)$, we can compute the probability

$$P(+++|xxx) = |\langle +x,+x,+x|GHZ\rangle|^2 = \left|\frac{1}{4}+\frac{1}{4}\right|^2 = \frac{1}{4}\,.$$

The other probabilities can be verified in a similar manner.

(2) From the statistics, we can take into account the outcomes with nonzero probabilities, which are $P(+++|xxx)$, $P(+--|xxx)$, $P(-+-|xxx)$ and $P(--+|xxx)$. By multiplying the outcomes of the three measurements for each of these four probabilities, we find that the results are always $+1$. For example, for $P(+--|xxx)$, we have

$$r_A(x)\,r_B(x)\,r_C(x) = (+1)(-1)(-1) = +1\,.$$

Thus we see that Equation (6.4) is valid with certainty for the measurement X-X-X.

Solution 6.2.

(1) We can verify these probabilities by using the statistics in Equation (6.5):

$$P(+|x) = P(+++|xxx) + P(+--|xxx) = \frac{1}{2}\,,$$

$$P(-|x) = P(-+-|xxx) + P(--+|xxx) = \frac{1}{2}\,.$$

(2) These probabilities can by calculated using the statistics in Equation (6.5). For example,

$$P(++|xx) = P(+++|xxx) + P(++-|xxx) = \frac{1}{4}\,.$$

We can do the same for the other probabilities.

Solution 6.3.

(1) We can compute $P(+++|xyy)$ by first expanding the expression

$$
\begin{aligned}
&|+x,+y,+y\rangle \\
&= \left(\frac{1}{\sqrt{2}}(|H\rangle + |V\rangle)\right)\left(\frac{1}{\sqrt{2}}(|H\rangle + i|V\rangle)\right)\left(\frac{1}{\sqrt{2}}(|H\rangle + i|V\rangle)\right) \\
&= \frac{1}{2\sqrt{2}}(|HHH\rangle + i|HHV\rangle + i|HVH\rangle - |HVV\rangle \\
&\quad +|VHH\rangle + i|VHV\rangle + i|VVH\rangle - |VVV\rangle)\,.
\end{aligned}
$$

We can then compute the probability

$$P(+++|xyy) = |\langle +x,+y,+y|GHZ\rangle|^2 = \left|\frac{1}{4} - \frac{1}{4}\right|^2 = 0\,.$$

The other probabilities can be verified in a similar manner.

(2) We prove the first equality by considering only the outcomes with nonzero probabilities, which are $P(++-|xyy)$, $P(+-+|xyy)$, $P(-++|xyy)$ and $P(---|xyy)$. We observe that for $P(++-|xyy)$,

$$r_A(x)\,r_B(y)\,r_C(y) \;=\; (+1)(+1)(-1) \;=\; -1\,.$$

We can prove this for the other three probabilities, and repeat this procedure for $r_A(y)\,r_B(x)\,r_C(y)$ and $r_A(y)\,r_B(y)\,r_C(x)$.

Solution 6.4. Combining (6.4) and (6.8), we have

$$r_B(x)\,r_C(x) \;=\; -r_B(y)\,r_C(y)\,.$$

Combining (6.9) and (6.10) gives us

$$r_B(x)\,r_C(y) \;=\; r_B(y)\,r_C(x)\,.$$

Combining the above two equalities, we obtain

$$r_C^2(x) \;=\; -r_C^2(y)\,,$$

which is a contradiction since $r_C^2(x) = r_C^2(y) = 1$. Thus we conclude that no list (6.12) can fulfill the four conditions.

1ST
LOOK AT THAT! HE SURE IS VERY COHERENT IN THE SPEECH COMPETITION!
LOOKS LIKE WHEN IT COMES TO INTERACTION, HE LOSES ALL HIS COHERENCE.
ERR...MOVIE? DRINK?...UMM...

Chapter 7

Measurement and Decoherence

In classical physics, measurements have always played a passive role. We assume that everything around us has well-defined properties, regardless of whether they are being observed or measured. After all, the moon is there even if no one is looking at it, isn't it?

In quantum physics, can we say the same about measurement? Does everything already have a predefined state before we measure them? If we measure a photon's polarization to be horizontal, can we say that it was already horizontal before we measured it? We know that this is problematic because of the violation of Bell's inequality.

In this chapter, we start by introducing the concept of pre-measurement, followed by decoherence, and a short discussion on the "measurement problem" in quantum physics. We shall finish by briefly mentioning the notion of quantum computing.

7.1 Measurement and Entanglement

Very often, in actual measurements, we do not measure the desired property directly. Rather, we couple it to a *pointer*, which is another property that will actually be measured. For example, when we use an analog ammeter to measure current, what we measure is the deflection of a needle (the pointer) when current passes through the circuit, and not the current itself. In quantum physics, the same happens, but the coupling of two quantum systems leads to entanglement.

We first consider using a polarizing beam-splitter to measure a photon's polarization in Figure (7.1). This beam-splitter allows photons with a horizontal polarization $|H\rangle$ to be transmitted, corresponding to path x, whereas photons with a vertical polarization $|V\rangle$ are reflected, correspond-

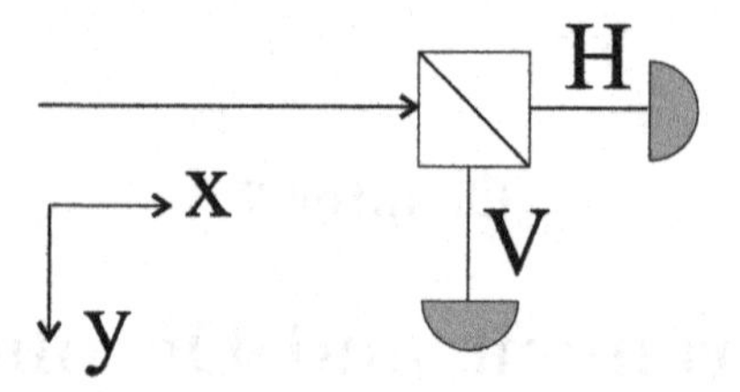

Fig. 7.1 How the polarization of a photon is measured.

ing to path y. We can thus deduce the photon's polarization by measuring the path that it takes.

Mathematically, this process can be written as

$$\big(\cos\theta|H\rangle + \sin\theta|V\rangle\big)|x\rangle \rightarrow \cos\theta|H\rangle|x\rangle + \sin\theta|V\rangle|y\rangle. \quad (7.1)$$

Here we see that the beam-splitter entangles the photon's polarization (the system) to its path (the pointer). From this, it is possible to predict the probabilities of measuring each polarization in either path.

Exercise 7.1. Fill up the table below.

Result of measurement	Probability
Polarization $|H\rangle$ in path $|x\rangle$	$\cos^2\theta$
Polarization $|V\rangle$ in path $|x\rangle$	
Polarization $|H\rangle$ in path $|y\rangle$	
Polarization $|V\rangle$ in path $|y\rangle$	

Note that (7.1) is a reversible transformation: at this stage no measurement has taken place yet. This is why this step is sometimes called pre-measurement: a coupling between the system and pointer such that

$$|H\rangle|x\rangle \longrightarrow |H\rangle|x\rangle$$
$$|V\rangle|x\rangle \longrightarrow |V\rangle|y\rangle$$

Given that $|H\rangle$ and $|V\rangle$ are orthogonal, we also want $|x\rangle$ and $|y\rangle$ to be orthogonal, because we want them to be perfectly distinguishable. This would allow us to deduce $|H\rangle$ or $|V\rangle$ from $|x\rangle$ or $|y\rangle$ with certainty.

Note that in this setup, $|H\rangle$ and $|V\rangle$ have to be orthogonal. Two non-orthogonal (and hence non-distinguishable) states cannot be coupled to two orthogonal states of the pointer, as can be verified in the following exercise.

Exercise 7.2. Suppose that such a pre-measurement is possible:

$$|H\rangle|x\rangle \xrightarrow{?} |H\rangle|x\rangle$$
$$|+\rangle|x\rangle \xrightarrow{?} |+\rangle|y\rangle$$

where $|+\rangle = \frac{1}{\sqrt{2}}(|H\rangle + |V\rangle)$. In this case, by measuring the pointer, we would be able to perfectly distinguish $|H\rangle$ from $|+\rangle$, even though these two states are not orthogonal. From what we have discussed so far in this book, convince yourself that such a transformation cannot be possible. This can be mathematically shown by proving that this transformation is not unitary, and is therefore forbidden. *Hint:* compute the scalar product of the state in the first line with the state in the second line, both before and after the transformation. If they are different, then the transformation is not unitary.

7.2 Decoherence

In the previous section, we have discussed pre-measurement using a polarizing beam-splitter. In actual fact, pre-measurement is always occurring around us, through the environment! This process is called *decoherence.* A degree of freedom is never decoupled from everything else: even if we can prepare a pure state, with time it will interact with the environment. In other words, part of the information initially contained in the system will diffuse into the environment. We can also say that the environment acquires information about the system.

Using the formalism of pre-measurement, we identify the environment as the pointer. We consider the following evolution in time:

$$\begin{aligned} |H\rangle|E\rangle &\xrightarrow{t} |H\rangle|E_H(t)\rangle\,, \\ |V\rangle|E\rangle &\xrightarrow{t} |V\rangle|E_V(t)\rangle\,. \end{aligned} \tag{7.2}$$

where $|E_H(t)\rangle$ and $|E_V(t)\rangle$ are two states of the environment, which are in general not orthogonal. We express these two states using an orthonormal basis $\{|E_1\rangle, |E_2\rangle\}$, by rewriting them as

$$\begin{aligned} |E_H(t)\rangle &= \sqrt{\lambda(t)}|E_1\rangle + \sqrt{1-\lambda(t)}|E_2\rangle\,, \\ |E_V(t)\rangle &= \sqrt{\lambda(t)}|E_2\rangle + \sqrt{1-\lambda(t)}|E_1\rangle\,. \end{aligned} \tag{7.3}$$

For simplicity of notation, we would not explicitly write the dependance on t of $|E_H\rangle$, $|E_V\rangle$ and λ.

Now, you may wonder, what does the parameter λ mean? How do $|E_H\rangle$ and $|E_V\rangle$ change with time? These are the questions we are going to study. It is convenient to denote the scalar product between the states of the environment as $\chi = \langle E_H|E_V\rangle$, this quantity plays an important role in what follows.

Exercise 7.3. Find an expression for χ as a function of λ.

We know that initially, at $t = 0$, the environment is in the state $|E\rangle$. Thus we can write that $|E_H\rangle = |E_V\rangle = |E\rangle$ at $t = 0$.

Exercise 7.4. Find λ and hence χ at $t = 0$, and express $|E\rangle$ in the basis $\{|E_1\rangle, |E_2\rangle\}$.

As time progresses, the environment states, $|E_H\rangle$ and $|E_V\rangle$, will evolve from being in the same state to becoming increasingly different as they interact more with the system. Thus intuitively, we know that after a very long time, $|E_H\rangle$ and $|E_V\rangle$ will become orthogonal.

Exercise 7.5. Find λ and χ for this case.

Now we define a function of χ with respect to time, $\chi(t) = e^{-t/\tau}$, where τ is called the "decoherence time". Intuitively, if the interaction with the environment is strong, χ should decrease rapidly with time, which means that τ should be short.

Exercise 7.6. Plot this function. Verify that the values of χ at $t = 0$ and as $t \to \infty$ correspond to those in the previous two exercises.

We now have an idea of how the environment varies as it interacts with the system. We have also seen that χ is related to the degree of the system's interaction with the environment. Next, we shall investigate how decoherence affects the outcomes of measurements made on the system.

Suppose that the initial state of the system is $|\psi\rangle|E\rangle = \big(\cos\theta|H\rangle + \sin\theta|V\rangle\big)|E\rangle$. Through interaction with the environment, at time t, this state has evolved into

$$\begin{aligned} |\Psi\rangle|E\rangle &= \left(\sqrt{\lambda}\cos\theta|H\rangle + \sqrt{1-\lambda}\sin\theta|V\rangle\right)|E_1\rangle \\ &\quad + \left(\sqrt{1-\lambda}\cos\theta|H\rangle + \sqrt{\lambda}\sin\theta|V\rangle\right)|E_2\rangle \\ &\equiv |\psi_1\rangle|E_1\rangle + |\psi_2\rangle|E_2\rangle\,. \end{aligned} \tag{7.4}$$

Exercise 7.7. Verify Equation (7.4).

Suppose that at time t, we choose to measure the polarization using the basis $|+\alpha\rangle = \cos\alpha|H\rangle + \sin\alpha|V\rangle$ and $|-\alpha\rangle = \cos\alpha|V\rangle - \sin\alpha|H\rangle$. Using Equation (7.4), it can be shown that the probabilities of obtaining each outcome are

$$\begin{aligned} P(+\alpha|\theta,\lambda) &= \tfrac{1}{2}\left(1 + \cos 2\theta\cos 2\alpha + \chi\sin 2\theta\sin 2\alpha\right), \\ P(-\alpha|\theta,\lambda) &= \tfrac{1}{2}\left(1 - \cos 2\theta\cos 2\alpha - \chi\sin 2\theta\sin 2\alpha\right). \end{aligned} \tag{7.5}$$

Hence we can derive the probabilities of each outcome as functions of θ, α and χ.

Exercise 7.8.

(1) Verify that the statistics are

$$\begin{aligned} P(\pm\alpha|\theta,\lambda) &= P(\pm\alpha|\psi_1) + P(\pm\alpha|\psi_2) \\ &= \left|\langle\pm\alpha|\psi_1\rangle\right|^2 + \left|\langle\pm\alpha|\psi_2\rangle\right|^2. \end{aligned} \tag{7.6}$$

(2) Compute explicitly

$$P(+\alpha|\theta,\lambda) = \left|\sqrt{\lambda}\cos\theta\cos\alpha + \sqrt{1-\lambda}\sin\theta\sin\alpha\right|^2 \tag{7.7}$$

$$+\left|\sqrt{1-\lambda}\cos\theta\cos\alpha + \sqrt{\lambda}\sin\theta\sin\alpha\right|^2 \tag{7.8}$$

$$= \frac{1}{2}\left(1 + \cos 2\theta\cos 2\alpha + \chi\sin 2\theta\sin 2\alpha\right) \tag{7.9}$$

where for the last line we used the trigonometric identities $\cos^2 x = \frac{1}{2}(1+\cos 2x)$ and $2\sin x\cos x = \sin 2x$. Notice that the only term here that depends on the environment is χ.

(3) Find a similar expression for $P(-\alpha|\theta,\lambda)$, and verify that $P(-\alpha|\theta,\lambda) = 1 - P(+\alpha|\theta,\lambda)$.

(4) Write explicitly $P(+\alpha|\theta,\lambda)$ and $P(-\alpha|\theta,\lambda)$ for $\alpha = \frac{\pi}{4}$, and plot these probabilities as functions of θ for $\chi = 1$, $\chi = \frac{1}{2}$ and $\chi = 0$.

If one plots the probabilities of the polarization measurements as a function of θ, as was suggested in the previous exercise for a particular case, one obtains sinusoidal functions with varying amplitudes depending on χ. When $\chi = 1$, we know that this corresponds to the initial state of the system, without any decoherence. This case corresponds to maximum amplitude, which means that interference effects are the most observable. On the other extreme, $\chi = 0$ corresponds to constant probabilities, and hence no interference effects. Thus, χ is related to how observable interference effects are.

The *visibility* of interference fringes is defined as:

$$V = \frac{P_{max} - P_{min}}{P_{max} + P_{min}}. \tag{7.10}$$

Here, P_{max} is the maximum probability for all values of θ; we write it as $P_{max} = \max_\theta \left[P(\pm\alpha|\theta, \lambda)\right]$. Similarly, $P_{min} = \min_\theta \left[P(\pm\alpha|\theta, \lambda)\right]$. In our case, it can be shown that $V = \chi$.

Exercise 7.9.

(1) Verify that $V = \chi$.

(2) From the previous plots of probability, the visibility also has a graphical meaning. Verify that $P_{max} = \frac{1+V}{2}$ and $P_{min} = \frac{1-V}{2}$. How do you read the visibility from the plots?

7.3 Summary

In pre-measurement, the desired property of a system is coupled to a pointer, leading to entanglement for quantum states. Decoherence is a form of pre-measurement, in which a system interacts with the environment. This causes the system and environment to evolve with time, resulting in a decrease of the visibility.

7.4 The Broader View

7.4.1 *The measurement problem*

We have devoted a full chapter to a somewhat detailed description of the measurement process in quantum physics. The reason is that the notion of measurement is not as trivial as in classical physics. One tends to think of measurement as a procedure that just reveals pre-existing properties; however, we have stressed many times in previous chapters that quantum phenomena cannot be ascribed to pre-existing properties. So, what does a measurement actually do? There is no agreement on this question; thus it has become customary to speak of the *measurement problem*.

The core of the issue seems to be the following: quantum physics itself has no sharp recipe to decide what is a measurement device. We have seen it in this chapter with the notion of the pointer: the pointer is another physical system that couples to the system to be measured and reads its information. In the text, we stopped there, but one might arguably ask: how do I measure the pointer now? You need a pointer for the pointer... If one pushes the formalism to its limit, even the observer can be treated as yet another pointer. This would give the following description of the measurement process: first, the system has been prepared in a state, but neither the detector nor the observer are correlated to it; then, in the second step, the detector correlates to the system; in the third step, the observer correlates to the detector. Formally:

$$\begin{aligned}
&(\cos\theta|H\rangle + \sin\theta|V\rangle)|?\rangle_{\text{det}}|?\rangle_{\text{obs}} \\
&\to (\cos\theta|H\rangle|\text{“}H\text{”}\rangle_{\text{det}} + \sin\theta|V\rangle|\text{“}V\text{”}\rangle_{\text{det}})|?\rangle_{\text{obs}} \\
&\to \cos\theta|H\rangle|\text{“}H\text{”}\rangle_{\text{det}}|\text{“}H\text{”}\rangle_{\text{obs}} + \sin\theta|V\rangle|\text{“}V\text{”}\rangle_{\text{det}}|\text{“}V\text{”}\rangle_{\text{obs}}\,.
\end{aligned}$$

The last line is a tripartite entangled state, very similar to the GHZ state of chapter 6.

Now, operationally, everything is fine: at any stage, you can just square the coefficients and recover the usual probabilities. So, the statistics will be the same independent of whether you choose the boundary between pre-measurement and actual measurement (incidentally, this boundary is often referred to as *Heisenberg's cut*).

But this nice feature hides a deep problem: does the last line correspond to something real? Is it really the case that, ultimately, no measurement ever “happens”, that one just has entanglement developing between more and more parties, including ourselves? This is the standpoint of the radical solution to the measurement problem proposed by Everett in 1957 and called the *many-worlds interpretation*. The daring elegance of this solution meets a problem: what is the sense of probabilities, of squaring coefficients? Why do we observe probabilities in the first place? In spite of several attempts by the proponents of this interpretation, no satisfactory answer has been given to this question.

The many-worlds interpretation just mentioned is rather radical and is not the most commonly adopted. The *orthodox interpretation* is the one we adopted in this book: at some point, a measurement really takes place. In this view, the notion of probabilities makes perfect sense, but of course one is assuming that there is a *real* Heisenberg's cut. At the moment of writing, nobody has any clue of what could define such a cut. Moreover, a measurement apparatus is made of atoms and atoms obey quantum physics: why should many atoms suddenly lose their quantum character?

There are several other interpretations; for the purpose of this text, let us mention only a group of *mechanistic interpretations*, the most famous and developed of which is called Bohmian mechanics. In Bohmian mechanics, everything is pre-established as in classical physics: in order to achieve this, one is obliged to postulate the existence of an all-pervasive "quantum potential", a sort of wave that carries no energy and has to change instantaneously everywhere when a measurement is made — this is a strongly non-local hidden variable, the non-locality being needed to justify the violation of Bell's inequality and similar phenomena. In Bohmian mechanics, measurement is not a problem because everything is deterministic. However, this interpretation is very controversial because it introduces physical objects that are in principle unobservable.

Let us summarize. The orthodox interpretation has a measurement problem that could be solved by discovering a real Heisenberg's cut, which however is problematic to define. The many worlds interpretation removes the measurement problem by saying that no real measurement ever takes place, but finds it difficult to give a meaning to our observation of probabilities. The Bohmian interpretation solves the measurement problem by recovering determinism, at the price however of postulating the existence of a strongly non-local and unobservable physical object. If you were feeling that you had not fully grasped quantum physics yet, now you know why: nobody really has!

7.4.2 *Quantum computing*

We have just seen that decoherence, i.e. interaction with the environment, is an effective explanation of the fact that we don't observe quantum interference of large objects. This very fact is also an important nuisance in the dream of building a quantum computer one day. We do not have space here to discuss this topic in depth, but it is useful to have some idea of what is at stake.

You have to first forget all the familiar features of a computer: the screen, keyboard etc., and focus on the *computations* that are on-going in the processor. Ultimately, any such computation takes a long list of zeros and ones as an input, and outputs another list of zeros and ones. But what is a zero or a one? In your computer, these are probably two different levels of voltages, or two orientations of a magnetic domain. In any case, the *logical values* "0" and "1" are coded into two states of a physical system.

Now, one may think of encoding the logical values in two quantum states $|0\rangle$ and $|1\rangle$. If one does this, the physical system that codes for the bit can also form superposition states $\alpha|0\rangle + \beta|1\rangle$, or be entangled with other systems. These operations have no analog in a computer operating on classical coding. Thus in principle, one can do more operations with quantum coding than with classical coding. The question is, is this useful? In 1994, Peter Shor showed that the answer is yes: the possibility of accessing superposition and entanglement can significantly speed up some computations. This means that a given computation would require *far fewer elementary steps*, it is not a question here of *how much time* each elementary step will take. In other words, the promise is not that quantum physics can easily produce terahertz processors; rather, if you have a classical megahertz processor and a quantum megahertz processor, the latter will perform some computations much faster than the former.

More than fifteen years after Shor, nobody has found any intuitive explanation of the power of quantum computing. One often hears the popular explanation: "a quantum computer would compute all possible results at the same time, because of superposition". This explanation is certainly *not* the heart of the matter: if it were, all possible computations would be sped up; on the contrary, it is known that only some specific computations are and many others are definitely not!

Now, in order to perform the type of computations that Shor envisaged, one needs to control thousands of quantum systems, to manipulate and measure each of them individually, and to entangle any two of them. This is a formidable experimental challenge: while nice demonstrations of such control have been made, they only involved a few quantum systems (approximately ten, the number varies with the technologies), and it is not clear how to scale the design up (the often-mentioned *scalability* issue). On top of this, the evolution of all the thousands of systems needs to be under control: decoherence here plays a negative role, because it is unwanted interaction with an environment that we don't control.

Can decoherence be harnessed? Or maybe even turned to an advantage by clever coding? Can one do meaningful computations with fewer systems than initially thought? All these questions are still being debated among researchers at the moment of writing. What we can say with assurance is, nobody has yet found a fully viable way to build a quantum computer.

7.5 References and Further Reading

Easy reading:

- On decoherence: W.H. Zurek, arXiv:quant-ph/0306072v1 (the updated version of an article published in Physics Today, 1991)
- On quantum computers: S. Lloyd, Scientific American **273**, 140 (1995); S. Aaronson, Scientific American 298, 62 (2008)

Resources:

- Online review by Guido Bacciagaluppi: http://plato.stanford.edu/entries/qm-decoherence/
- Review article: M. Schlosshauer, Rev. Mod. Phys. **76**, 1267 (2004).

Suggestion for projects:

- M. Brune et al., Phys. Rev. Lett. **77**, 4887 (1996).

7.6 Solutions to the Exercises

Solution 7.1. To calculate the probability of the photon with polarization $|H\rangle$ to be in path $|x\rangle$, we square the scalar product of $\big(\cos\theta|H\rangle|x\rangle + \sin\theta|V\rangle|y\rangle\big)$ and $|H\rangle|x\rangle$, which gives us $\cos^2\theta$. We can determine all the other probabilities using this method.

Result of measurement	Probability
Polarization $\lvert H\rangle$ in path $\lvert x\rangle$	$\cos^2\theta$
Polarization $\lvert V\rangle$ in path $\lvert x\rangle$	0
Polarization $\lvert H\rangle$ in path $\lvert y\rangle$	0
Polarization $\lvert V\rangle$ in path $\lvert y\rangle$	$\sin^2\theta$

Solution 7.2. Let us first compute the scalar product of the states before the transformation. Since

$$|+\rangle|x\rangle = \frac{1}{\sqrt{2}}\left(|H\rangle|x\rangle + |V\rangle|x\rangle\right) ,$$

we have

$$\langle H, x|+, x\rangle = \frac{1}{\sqrt{2}}\,.$$

where $|H, x\rangle$ stands for $|H\rangle \otimes |x\rangle$; similarly for $|+, x\rangle$.

Now we compute the scalar product of the states after the transformation. Since

$$|+\rangle|y\rangle = \frac{1}{\sqrt{2}}\left(|H\rangle|y\rangle + |V\rangle|y\rangle\right),$$

thus

$$\langle H, x|+, y\rangle = 0\,.$$

Since the two scalar products are not equal, this transformation is not unitary and is hence forbidden.

Solution 7.3. $\chi = 2\sqrt{\lambda(1-\lambda)}$.

Solution 7.4. At $t = 0$, $|E_H\rangle = |E_V\rangle$. Thus

$$\begin{aligned}\sqrt{\lambda}|E_1\rangle + \sqrt{1-\lambda}|E_2\rangle &= \sqrt{\lambda}|E_2\rangle + \sqrt{1-\lambda}|E_1\rangle\,,\\ \sqrt{\lambda} &= \sqrt{1-\lambda}\,.\end{aligned}$$

Hence we see that $\lambda(t = 0) = \frac{1}{2}$. Using this value of λ, we obtain $\chi = 1$. This can also be deduced by noting that $|E_V\rangle$ is the same as $|E_H\rangle$. Using equation (7.3), we can then express $|E\rangle$ in the basis $\{|E_1\rangle, |E_2\rangle\}$ as follows:

$$|E\rangle = \sqrt{\frac{1}{2}}|E_1\rangle + \sqrt{1-\frac{1}{2}}|E_2\rangle = \frac{1}{\sqrt{2}}\left(|E_1\rangle + |E_2\rangle\right).$$

Observe that the environment has an equal probability of being in either of the two orthogonal states $|E_1\rangle$ and $|E_2\rangle$.

Solution 7.5. If $|E_H\rangle$ and $|E_V\rangle$ are orthogonal, this means that $\chi = 0$, which in turn corresponds to $\lambda = 1$.

Solution 7.6. From the graph, we can see that at $t = 0$, $\chi = 1$, and in the limit as $t \to \infty$, $\chi \to 0$. This corresponds to the previous two exercises.

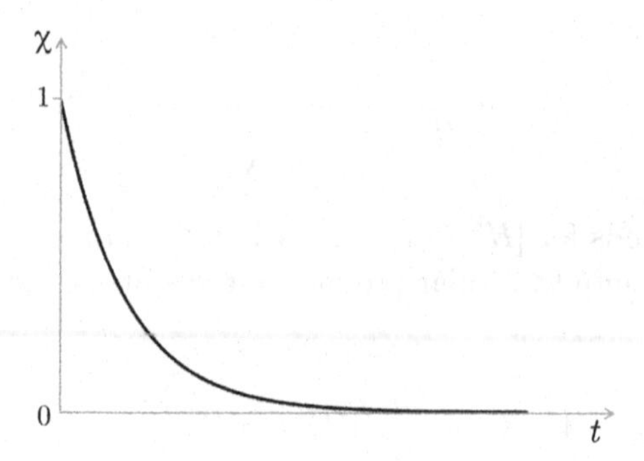

Fig. 7.2 Plot of χ with respect to time t.

Solution 7.7. We can transform the initial state to obtain:

$$\begin{aligned}
|\psi\rangle|E\rangle &= \big(\cos\theta|H\rangle + \sin\theta|V\rangle\big)|E\rangle \\
&\longrightarrow \cos\theta|H\rangle|E_H\rangle + \sin\theta|V\rangle|E_V\rangle \\
&= \cos\theta|H\rangle\left(\sqrt{\lambda}|E_1\rangle + \sqrt{1-\lambda}|E_2\rangle\right) \\
&\quad + \sin\theta|V\rangle\left(\sqrt{\lambda}|E_2\rangle + \sqrt{1-\lambda}|E_1\rangle\right).
\end{aligned}$$

Rearranging this, we get

$$\begin{aligned}
|\Psi\rangle|E\rangle = &\left(\sqrt{\lambda}\cos\theta|H\rangle + \sqrt{1-\lambda}\sin\theta|V\rangle\right)|E_1\rangle \\
&+\left(\sqrt{1-\lambda}\cos\theta|H\rangle + \sqrt{\lambda}\sin\theta|V\rangle\right)|E_2\rangle \\
\equiv &\ |\psi_1\rangle|E_1\rangle + |\psi_2\rangle|E_2\rangle,
\end{aligned}$$

where $|\psi_1\rangle = \sqrt{\lambda}\cos\theta|H\rangle + \sqrt{1-\lambda}\sin\theta|V\rangle$, and $|\psi_2\rangle = \sqrt{1-\lambda}\cos\theta|H\rangle + \sqrt{\lambda}\sin\theta|V\rangle$.

Solution 7.8.

(1) As the system interacts with the environment, the original state $|\psi\rangle$ evolves to a state having probabilities of being in either $|\psi_1\rangle$ or $|\psi_2\rangle$. Thus to find $P(\pm\alpha|\theta,\lambda)$, we find the individual probabilities for the cases in which the state is $|\psi_1\rangle$ and $|\psi_2\rangle$. The total probability is obtained by summing these up.

(2) We can expand the expression from the previous question to obtain

$$\begin{aligned}
&P(+\alpha|\theta,\lambda)\\
&=\left|\langle+\alpha|\psi_1\rangle\right|^2+\left|\langle+\alpha|\psi_2\rangle\right|^2\\
&=\left|\sqrt{\lambda}\cos\theta\cos\alpha+\sqrt{1-\lambda}\sin\theta\sin\alpha\right|^2\\
&\quad+\left|\sqrt{1-\lambda}\cos\theta\cos\alpha+\sqrt{\lambda}\sin\theta\sin\alpha\right|^2\\
&=\sin^2\theta\sin^2\alpha+\cos^2\theta\cos^2\alpha+4\sqrt{\lambda(1-\lambda)}\sin\theta\cos\theta\sin\alpha\cos\alpha\\
&=\frac{1}{4}(1-\cos2\theta)(1-\cos2\alpha)+\frac{1}{4}(1+\cos2\theta)(1+\cos2\alpha)\\
&\quad+\frac{1}{2}\chi\sin2\theta\sin2\alpha\\
&=\frac{1}{2}(1+\cos2\theta\cos2\alpha+\chi\sin2\theta\sin2\alpha)\,.
\end{aligned}$$

(3) We use the same method as above, except that the state is now $|-\alpha\rangle$ instead of $|+\alpha\rangle$.

$$\begin{aligned}
&P(-\alpha|\theta,\lambda)\\
&=\left|\langle-\alpha|\psi_1\rangle\right|^2+\left|\langle-\alpha|\psi_2\rangle\right|^2\\
&=\left|\sqrt{1-\lambda}\sin\theta\cos\alpha-\sqrt{\lambda}\cos\theta\sin\alpha\right|^2\\
&\quad+\left|\sqrt{\lambda}\sin\theta\cos\alpha-\sqrt{1-\lambda}\cos\theta\sin\alpha\right|^2\\
&=\sin^2\theta\cos^2\alpha+\cos^2\theta\sin^2\alpha-4\sqrt{\lambda(1-\lambda)}\sin\theta\cos\theta\sin\alpha\cos\alpha\\
&=\frac{1}{4}(1-\cos2\theta)(1+\cos2\alpha)+\frac{1}{4}(1+\cos2\theta)(1-\cos2\alpha)\\
&\quad-\frac{1}{2}\chi\sin2\theta\sin2\alpha\\
&=\frac{1}{2}(1-\cos2\theta\cos2\alpha-\chi\sin2\theta\sin2\alpha)\\
&=1-P(+\alpha|\theta,\lambda)\,.
\end{aligned}$$

(4) Substituting $\alpha=\frac{\pi}{4}$,

$$P(+\alpha|\theta,\lambda)=\frac{1}{2}(1+\chi\sin2\theta)\,,$$

$$P(-\alpha|\theta,\lambda)=\frac{1}{2}(1-\chi\sin2\theta)\,.$$

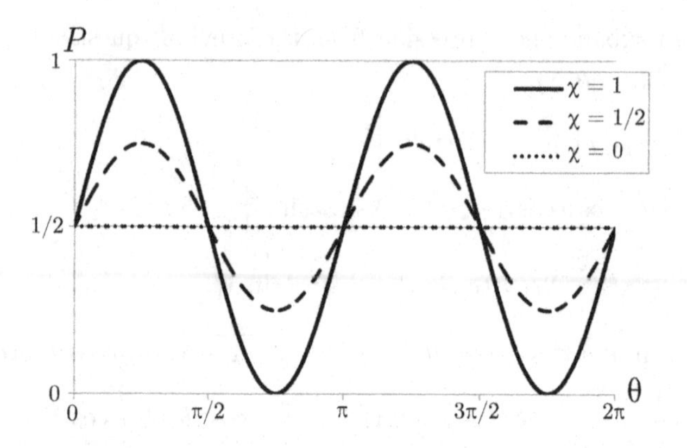

Fig. 7.3 Graphs of $P(+\alpha|\theta, \lambda)$

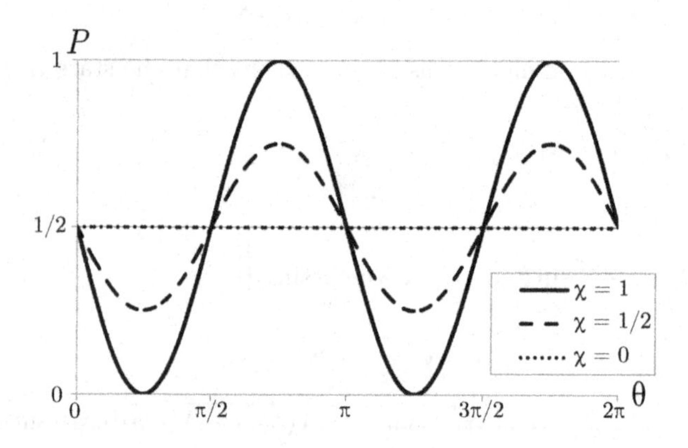

Fig. 7.4 Graphs of $P(-\alpha|\theta, \lambda)$

Solution 7.9.

(1) The maximum and minimum values of a sine function are +1 and −1 respectively. Thus

$$P_{max} = \frac{1+\chi}{2},$$

$$P_{min} = \frac{1-\chi}{2},$$

$$V = \frac{P_{max} - P_{min}}{P_{max} + P_{min}} = \chi\,.$$

(2) It is easy to verify that $P_{max}+P_{min}=1$. We can substitute this into the expression for V and solve it to obtain $P_{max}=\frac{1+V}{2}$ and $P_{min}=\frac{1-V}{2}$. To read the visibility from the plots, we take

$$P_{max}-P_{min}=V.$$

Thus we see that V is just twice the amplitude of the probability plots.

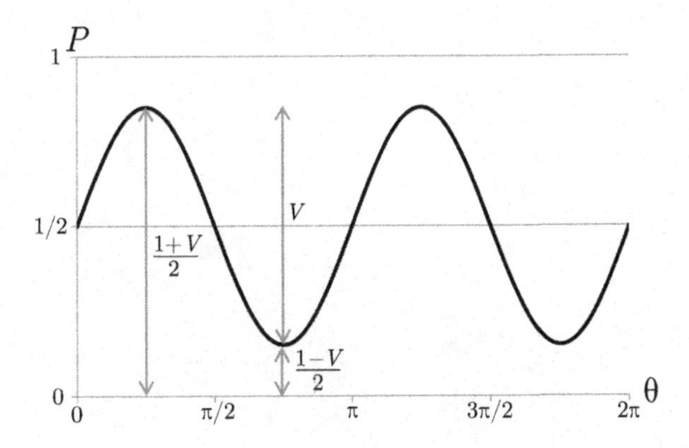

Fig. 7.5 Graphical determination of visibility

PART 3
Beyond the Six Pieces

Chapter 8

Other Two-Level Systems

We have introduced quantum physics using polarization and we have presented all the previous phenomena using that language. In this chapter, we briefly review other degrees of freedom that share with polarization the fact of being *two-level systems*, i.e. systems for which only two states can be perfectly distinguished. This chapter will show how the same formalism can be applied to describe a larger variety of phenomena.

8.1 Spin $\frac{1}{2}$

In many textbooks, the typical example of a two-level system is the *spin* $\frac{1}{2}$, a property of many elementary particles, notably protons, neutrons and electrons.

8.1.1 *Spins as intrinsic magnetic moments*

The spin is a magnetic moment, like the needle of a compass. Its state is defined by a direction, the direction in which the needle points. The most typical dynamic behavior of spins is observed when they are placed in a magnetic field: in the absence of dissipation, they rotate around the axis defined by the field with constant angular velocity.

The name "spin" misleadingly suggests that something is spinning. In order to understand the origin of this name, we have to embark in a sketchy historical remark. The existence of magnetic effects for electrons was predicted in the very early days of quantum physics: indeed, since electrons are charged particles, they generate a magnetic moment by rotating around the nucleus. However, this mechanism alone could not explain all the observed magnetic effects: an additional magnetic moment was needed. Physicists started looking for another mechanism that would explain it, and surmised that it could be due to the spinning of the electron around its axis; thus the word "spin". The name stuck, even though it became clear very soon that this explanation was completely inadequate — the reader of this book should know by now that it is sometimes impossible to find classical mechanisms to explain quantum physics.

In summary, the spin is an intrinsic magnetic moment. It does not derive from rotating charges or other dynamical effects: it is a purely quantum property of some particles. As for the number $\frac{1}{2}$, it is related to the number of orthogonal states: in general, a spin S has $d = 2S + 1$ orthogonal states. For the reason why S, instead of d, is chosen to label spins, we refer the reader to any standard textbook. For the purpose of this text, just keep in mind that a spin $\frac{1}{2}$ has $d = 2 \times \frac{1}{2} + 1 = 2$ states and is therefore a two-level system.

8.1.2 *States of spins*

8.1.2.1 *States of one spin*

The mathematical formalism is exactly the same as for polarization; only the interpretation of the states changes. The most common basis, which was $\{|H\rangle, |V\rangle\}$ for polarization, is usually written $\{|\uparrow\rangle, |\downarrow\rangle\}$ or $\{|+\hat{z}\rangle, |-\hat{z}\rangle\}$, and is interpreted as the spin pointing upwards or downwards in the $\hat{z}$ direction — of course, the choice of *which* direction in space is the $\hat{z}$ direction is completely arbitrary, just as the choice of what H and V are for polarization.

Already at this stage, one finds an issue that creates some confusion for beginners: note that orthogonal quantum states of spins are associated to opposite, and not to orthogonal, directions in space! There is nothing wrong with this, if you remember that the quantum state describes the properties of a system: the statement just means that two states of spin can be perfectly discriminated if and only if they indicate opposite directions. Still, one has to get accustomed to these subtleties, and this is one of the reasons for our choice of polarization as the degree of freedom in this book.

One consequence of this fact is the following, which we give without further proof: a spin that points in the direction $\cos\theta\hat{z} + \sin\theta\hat{x}$ is described by the state $|\psi(\theta)\rangle = \cos\frac{\theta}{2}|+\hat{z}\rangle + \sin\frac{\theta}{2}|-\hat{z}\rangle$. When one adds the third direction, complex numbers must be used: the spin that points in the direction $\hat{n}(\theta,\varphi) = \cos\theta\hat{z} + \sin\theta[\cos\varphi\hat{x} + \sin\varphi\hat{y}]$ is

$$|+\hat{n}\rangle \equiv |\psi(\theta,\varphi)\rangle = \cos\frac{\theta}{2}|+\hat{z}\rangle + e^{i\varphi}\sin\frac{\theta}{2}|-\hat{z}\rangle\,. \tag{8.1}$$

Exercise 8.1.

(1) Write down explicitly the four states $|+\hat{x}\rangle$, $|-\hat{x}\rangle$, $|+\hat{y}\rangle$ and $|-\hat{y}\rangle$. Verify that $\langle +\hat{x}|-\hat{x}\rangle = \langle +\hat{y}|-\hat{y}\rangle = 0$. Important remark: when complex numbers are involved, the scalar product is defined as follows: if $|\psi\rangle = a_1|+\hat{z}\rangle + a_2|-\hat{z}\rangle$ and $|\phi\rangle = b_1|+\hat{z}\rangle + b_2|-\hat{z}\rangle$, then $\langle\psi_1|\psi_2\rangle = a_1^*b_1 + a_2^*b_2$, where the "*" denotes complex conjugation.
(2) Write down the state $|-\hat{n}\rangle$ orthogonal to $|+\hat{n}\rangle$. Hint: which angles (θ',φ') define the direction $\hat{n}(\theta',\varphi') = -\hat{n}(\theta,\varphi)$?

For a beam of neutral particles, a measurement of spin along a given direction is made by a device called the *Stern-Gerlach magnet.* This device is described in all textbooks of quantum physics and we refer to those for all the details: the reader will see that it is the analog for spins of the polarizing beam-splitter that we have introduced in this book.

8.1.2.2 *Entanglement of two spins*

Surely the reader of this book has some elementary knowledge of the periodic table of elements and has therefore learnt how different elements appear: the electrons around the nucleus must occupy some well-defined *orbitals*, and each orbital can accommodate at most two electrons. Why? Because of Pauli's exclusion principle, two electrons cannot have the same state, so if they are in the same orbital they must have opposite spins: one must be "up", the other must be "down".

Now, this rule is already so surprising and new that students are overwhelmed and invariably fail to ask an obvious question: *along which direction* are the two spins opposite? Maybe it's good that they don't ask this during a first presentation, because the answer involves entanglement! The actual state is the so-called *singlet state*:

$$|\Psi^-\rangle = \frac{1}{\sqrt{2}}\left(|+\hat{z}\rangle|-\hat{z}\rangle - |-\hat{z}\rangle|+\hat{z}\rangle\right)\,. \tag{8.2}$$

As written, it seems that the spins are opposite in the $\hat{z}$ direction. But this cannot be the whole story, because the $\hat{z}$ direction is our arbitrary choice: how can the electrons know about it? In fact, there is no issue, thanks to entanglement: the singlet state has the remarkable property that it looks the same in every basis. Indeed, for any direction $\hat{n}$ of our choice, the expression

$$\frac{1}{\sqrt{2}}(|+\hat{n}\rangle|-\hat{n}\rangle - |-\hat{n}\rangle|+\hat{n}\rangle) \tag{8.3}$$

represents *the same state*. So, the spins of the two electrons in the same orbital point in opposite directions ... for all directions!

Exercise 8.2. Replace the states $|\pm\hat{n}\rangle$ in Equation (8.3) using Equation (8.1) and the result found in Exercise 8.1. Show that this state is identical to the one defined in (8.2), up to a multiplicative complex number $-e^{i\varphi}$ that does not play any role.

We have just found that there is entanglement of spins in virtually any atom, as soon as an orbital is occupied by two electrons. And an atom is just the simplest example: if one moves to several atoms, for instance in a solid, one can find many-electron entanglement, similar to the one discussed for the GHZ argument. However, the fact that there is entanglement in the electrons that compose the pages of this book does not make it any simpler to observe this entanglement. For instance, nobody knows how to extract two entangled electrons, send them far apart and check for a violation of Bell's inequality by measuring their spins.

In summary, entanglement of spins is ubiquitous; the problem is how to detect it. Interestingly, John Bell derived his inequality using the language of spins, not polarization, and thinking precisely in terms of the singlet state. But the experiments were finally done with photons.

8.2 Selected States

Polarization and spin $\frac{1}{2}$ are, so to say, natural two-level systems. One can obtain effective two-level systems out of more-level systems, if only two states and their superpositions are involved in an experiment. This procedure may seem artificial, but in actual fact some of the most meaningful examples of two-level systems are obtained precisely in this way. Here we briefly discuss two of them: two possible paths for a propagating particle and two selected energy levels of an electron in an atom.

8.2.1 *Two paths: interferometry*

Consider a source emitting a beam of light. In the previous chapters, we have studied the polarization of the beam. But actually, with light, something even simpler can be done: one can split the beam into two using a semi-transparent mirror, also called a *beam-splitter*. This effect is well-known from everyday life: just look at a window pane or through some spectacles, and you will see that much light is transmitted but part of it is also reflected.

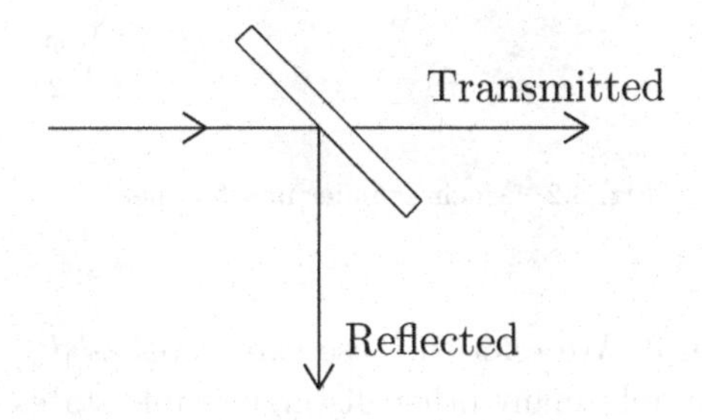

Fig. 8.1 Beam-splitter.

Then one can ask the same question that we asked at the beginning for polarization: how does one describe these partial reflections in terms of single photons? What does each photon do? The simplest experiment (Figure 8.1) shows that each photon is found either on the transmitted or on the reflected path: one never finds a "half-photon" in each path. This seems to settle the issue: each photon is either transmitted or reflected. But the reality is more interesting.

In order to understand this better, we have to complicate the setup (Figure 8.2). Using mirrors, one can recombine the initially split beams on another beam-splitter; prior to that, a piece of matter with larger refractive index can be placed in one of the paths to make it slightly longer. Such a setup is called a *Mach-Zehnder interferometer*. Let us do the calculations for this setup.

First, we have to identify the distinguishable states. We can distinguish a photon propagating along the direction $\hat{x}$ from one propagating

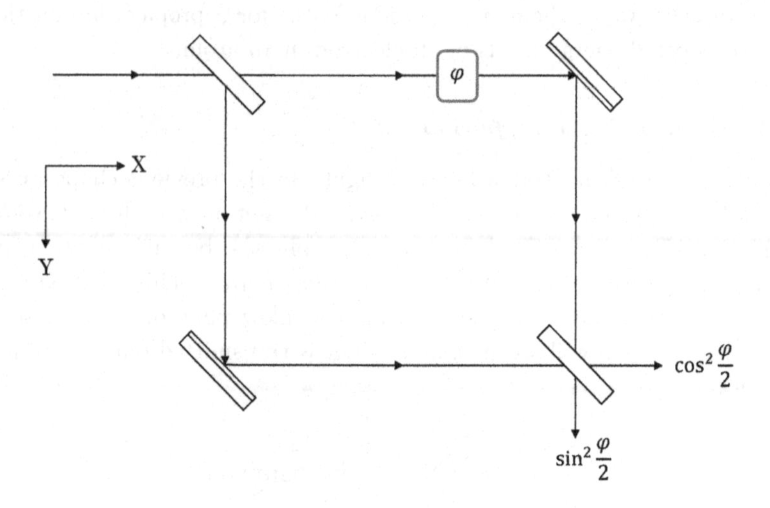

Fig. 8.2 Mach-Zehnder interferometer.

along the direction $\hat{y}$. We refer to these two states as $|k_x\rangle$ and $|k_y\rangle$. Note that there are infinitely many other distinguishable states describing photons that propagate in other directions, but these states do not enter into the description of the Mach-Zehnder interferometer, so we do not have to consider them. In this sense, the setup is described by an effective two-level system. Second, we note that a beam-splitter acts as the following transformation:

$$\begin{aligned} |k_x\rangle &\longrightarrow \tfrac{1}{\sqrt{2}}(|k_x\rangle + i|k_y\rangle)\,, \\ |k_y\rangle &\longrightarrow \tfrac{1}{\sqrt{2}}(|k_y\rangle + i|k_x\rangle)\,. \end{aligned} \tag{8.4}$$

The presence of the imaginary i is needed to guarantee that the transformation is unitary. Lastly, we need to describe the delay that is inserted into path x. Its effect is simply $|k_x\rangle \longrightarrow e^{i\varphi}|k_x\rangle$, while $|k_y\rangle$ is obviously left unchanged.

Exercise 8.3. Verify that the transformation (8.4) is indeed unitary as written, and would not be unitary if the imaginary number i is removed from the right-hand side.

Having all the pieces, we can start with an incoming particle on path x and propagate it through the setup:

$$\begin{aligned}
|k_x\rangle &\overset{\text{BS}_1}{\longrightarrow} \frac{1}{\sqrt{2}}(|k_x\rangle + i|k_y\rangle) \\
&\overset{\text{delay}}{\longrightarrow} \frac{1}{\sqrt{2}}(e^{i\varphi}|k_x\rangle + i|k_y\rangle) \\
&\overset{\text{mirrors}}{\longrightarrow} \frac{1}{\sqrt{2}}(e^{i\varphi}|k_y\rangle + i|k_x\rangle) \\
&\overset{\text{BS}_2}{\longrightarrow} \frac{1}{\sqrt{2}}\left[\frac{e^{i\varphi}}{\sqrt{2}}(|k_y\rangle + i|k_x\rangle) + \frac{i}{\sqrt{2}}(|k_x\rangle + i|k_y\rangle)\right] \\
&= i\frac{e^{i\varphi}+1}{2}|k_x\rangle + \frac{e^{i\varphi}-1}{2}|k_y\rangle\,.
\end{aligned}$$

So, at the output of the interferometer, the probability of finding the photon propagating along $\hat{x}$ is

$$P(k_x) = \left|i\frac{e^{i\varphi}+1}{2}\right|^2 = \cos^2\frac{\varphi}{2}\,; \tag{8.5}$$

similarly, the probability of finding the photon propagating along $\hat{y}$ is

$$P(k_y) = \left|\frac{e^{i\varphi}-1}{2}\right|^2 = \sin^2\frac{\varphi}{2}\,. \tag{8.6}$$

We see that, for $\varphi = 0$, $P(k_x) = 1$: *all* the photons will propagate out along $\hat{x}$; while for $\varphi = \pi$, $P(k_y) = 1$: *all* the photons will propagate out along $\hat{y}$. But then, it is sufficient to change φ in one path to change the state of all the photons: therefore, necessarily, each photon collects the information available on both paths — in other words, each photon is *delocalized* in both paths.

This delocalization is quite a striking effect of quantum superposition; so much so that the Mach-Zehnder interferometer is often chosen as the simplest setup to demonstrate the quantum behavior of a single particle. We rather chose polarization for our main text for the following two reasons: first, the superposition of two polarization states defines another direction of polarization; on the contrary, the superposition of two path states does not define another path, but a delocalized particle. There is thus an additional layer of conceptual difficulty in interpreting the superposition states. Second, experiments with entanglement become more cumbersome to describe in terms of paths (though entanglement in paths is certainly possible and has been observed as well).

8.2.2 *Two energy levels*

Electrons that are bound around a nucleus in an atom can take only discrete values of energy; this well-known fact is the origin of the spectral lines observed in the emission and absorption of light, one of the phenomena that triggered the discovery of quantum physics. One needs the full quantum theory of position and momentum in order to predict these energies; this goes beyond the scope of this book and will only be dealt with superficially in the next chapter. Let us just accept that there are states $|E_1\rangle$, $|E_2\rangle$, $|E_3\rangle$ and so on associated to each of these discrete energy levels.

As spectral lines indicate, one can couple one energy level to another by shining light at the suitable frequency and polarization. So, again, one can consider an experiment in which only two levels are coupled; the others are of course there, but they never come into play. The two meaningful states are generically called $|g\rangle$ or "ground" for the state of lower energy, $|e\rangle$ or "excited" for the state of higher energy: yet another example of an effective two-level system.

But we are dealing with quantum systems, therefore any state of the form $c_g|g\rangle + c_e|e\rangle$ is also a possible state. What does such a state mean? It is a state of undefined energy, something for which we lack intuition, possibly even more than for spatially delocalized states. Once again, while the meaning of such states is hard to convey, their existence is largely vindicated by observations and carefully designed experiments. We do not have time to describe these ones here (see below for some hints); the reader who studies quantum physics further than this book will certainly meet them.

8.3 Summary

Two-level systems are systems for which only two states can be perfectly distinguished. Spin, an intrinsic magnetic property of electrons, is an example of such a two-level system. Spin is similar to polarization and can be entangled, but the detection of entangled electrons is still a challenge. Other two-level systems can be obtained from more-level systems. An interferometer consists of two possible paths for propagating particles, giving rise to delocalized particles that explore both paths. Two energy levels in an atom can be similarly coupled, resulting in a state of undefined energy. These systems exemplify the interesting phenomenon of entanglement.

8.4 The Broader View

8.4.1 *Two systems on a single "particle"*

This subsection is devoted to a remark that is somehow obvious, but rich in consequences. Consider the Mach-Zehnder interferometer: we have described it for photons, so one may ask, what about polarization? Clearly, we could neglect it, because neither the beam-splitters nor the mirrors affect the polarization. In other words, the polarization remains the same in the whole interferometer, as the polarization of the input beam.

But what if one adds, in the interferometer, a device that *does* change the polarization? For instance, suppose that in path $\hat{x}$, before the delay, one adds a plate that rotates the polarization. Then, one has to study both degrees of freedom: path and polarization. Even if there is only one photon, it is a composite system!

For the sake of the calculation, call $\hat{H}$ the initial polarization, and suppose that the rotation is $|H\rangle \rightarrow |\alpha\rangle = \cos\alpha|H\rangle + \sin\alpha|V\rangle$. Then, the calculation starts as follows:

$$
\begin{aligned}
|k_x\rangle|H\rangle &\stackrel{\text{BS}_1}{\longrightarrow} \frac{1}{\sqrt{2}}(|k_x\rangle + i|k_y\rangle)|H\rangle \\
&\stackrel{\text{rotation}}{\longrightarrow} \frac{1}{\sqrt{2}}(|k_x\rangle|\alpha\rangle + i|k_y\rangle|H\rangle) \\
&\stackrel{\text{delay}}{\longrightarrow} \ldots
\end{aligned}
$$

We leave it for those who are interested to complete the calculation, but by comparison with chapter 7, the reader should see what happens: entanglement with another degree of freedom, hence *decoherence*! The case is pretty obvious if the rotation is such that $|H\rangle$ is transformed into $|V\rangle$: then the polarization identifies the path and no interference will be observed.

Needless to say, the same effect can be observed using particles with spin instead of photons and rotating the spin on one path using a localized magnetic field; this is typically done with neutrons. Yet another variation: an interferometer for atoms, in which the energy level is modified in one of the paths. All these experiments have been performed; we give some references on the next page.

8.4.2 *Ramsey interferometry: pulsed NMR and atomic clocks*

For consistency with the structure of the present book, we have focused on the preparation and measurement of states and said very little about *evolution in time.* The purpose of this paragraph is to sketch a specific measurement procedure, generically known as *Ramsey interferometry*, that is widely used in different contexts. Ramsey interferometry is most simply described with spins $\frac{1}{2}$, which we thus do here.

Consider a spin in a constant magnetic field, initially aligned along the field (direction $\hat{z}$). Suppose that one is able to induce a fast rotation on the spin around the $\hat{x}$ axis. Then one can apply a fast rotation by ninety degrees, at the end of which the spin is found along direction $\hat{y}$. Now the spin feels the magnetic field again, but is no longer aligned with it: as it should, it starts precessing around the $\hat{z}$ axis, i.e. in the $(\hat{x}, \hat{y})$ plane, with angular velocity ω_B proportional to the intensity of the magnetic field. Some time τ later, another fast rotation around $\hat{x}$ is applied: where does the spin end up?

Of course, *it depends on* τ! If $\tau = 0$, then the two rotations are simply combined and the spin ends up along $-\hat{z}$. If $\omega_B\tau = \frac{\pi}{2}$, the spin has rotated from $\hat{y}$ to $\hat{x}$, so a rotation around $\hat{x}$ does not change it: the spin remains along $\hat{x}$. If $\omega_B\tau = \pi$, the spin has rotated from $\hat{y}$ to $-\hat{y}$, so a rotation around $\hat{x}$ brings it back to $\hat{z}$. And so on. The probability of finding the spin in the state $|-\hat{z}\rangle$ as a function of τ is not difficult to compute in general: its value is $P(\tau) = \cos^2 \frac{\omega_B\tau}{2}$.

Note the striking analogy with the Mach-Zehnder interferometer described earlier: the fast rotations around $\hat{x}$ play the role of the beam-splitters, the time τ during which the system is left precessing plays the role of the delay φ. The main difference is that in this case, the interference is between spin states, something much more abstract than the optical interference between different paths. This sequence is at the core of *nuclear magnetic resonance (NMR)*, a powerful spectroscopic technique best known for its applications in medical imagery.

Exactly the same procedure can be applied to energy states, in which the fast rotation is induced by a light pulse at a suitable frequency. This is the principle on which *atomic clocks* are built: since the frequency of oscillation ω_B between two energy levels is known very precisely, the measurement of $P(\tau)$ gives access to the time τ with the same high precision.

8.5 References and Further Reading

Easy reading:

- B.-G. Englert, M.O. Scully, H. Walther, Scientific American **271**, 86 (1994).

- Chapters 1-4 of: V. Scarani, *Quantum Physics: A First Encounter* (Oxford University Press, Oxford, 2006).

Resources:

- Chapters 5, 6, 11 and 12 of: R.P. Feynman, R.B. Leighton, M. Sands, *The Feynman Lectures on Physics*, Vol 3: Quantum mechanics (Addison Wesley, 2nd edition 2005).

Suggestions for projects:

- S. Dürr, T. Nonn, G. Rempe, Nature **395**, 33 (1998).
- P. Bertet et al., Nature **411**, 166 (2001).

8.6 Solutions to the Exercises

Solution 8.1.

(1) In spherical coordinates, the direction $+\hat{x}$ is given by $\theta = \frac{\pi}{2}$, $\phi = 0$; $-\hat{x}$ is given by $\theta = \frac{\pi}{2}$, $\phi = \pi$. Similarly, the directions $+\hat{y}$ and $-\hat{y}$ are given by $\theta = \frac{\pi}{2}$, $\phi = \frac{\pi}{2}$ and $\theta = \frac{\pi}{2}$, $\phi = \frac{3\pi}{2}$ respectively. Substituting these angles into Equation (8.1), we obtain

$$\begin{aligned}
|+\hat{x}\rangle &= \frac{\sqrt{2}}{2}\left(|+\hat{z}\rangle + |-\hat{z}\rangle\right), \\
|-\hat{x}\rangle &= \frac{\sqrt{2}}{2}\left(|+\hat{z}\rangle - |-\hat{z}\rangle\right), \\
|+\hat{y}\rangle &= \frac{\sqrt{2}}{2}\left(|+\hat{z}\rangle + i|-\hat{z}\rangle\right), \\
|-\hat{y}\rangle &= \frac{\sqrt{2}}{2}\left(|+\hat{z}\rangle - i|-\hat{z}\rangle\right).
\end{aligned}$$

From the above expressions, we can calculate the scalar products

$$\begin{aligned}
\langle +\hat{x}|-\hat{x}\rangle &= \frac{1}{2} - \frac{1}{2} = 0, \\
\langle +\hat{y}|-\hat{y}\rangle &= \frac{1}{2} + i^2\frac{1}{2} = 0.
\end{aligned}$$

(2) The angles (θ', φ') that define $\hat{n}(\theta', \varphi') = -\hat{n}(\theta, \varphi)$ in spherical coordinates are

$$\begin{aligned}
\theta' &= \pi - \theta, \\
\varphi' &= \varphi + \pi.
\end{aligned}$$

Substituting these angles into Equation (8.1), we obtain

$$|-\hat{n}\rangle = \sin\frac{\theta}{2}|+\hat{z}\rangle - e^{i\varphi}\cos\frac{\theta}{2}|-\hat{z}\rangle\,. \tag{8.7}$$

Solution 8.2. Substituting Equations (8.1) and (8.7) into (8.3), and using the trigonometric identity $\cos^2\frac{\theta}{2}+\sin^2\frac{\theta}{2}=1$, we obtain

$$\frac{1}{\sqrt{2}}\left(|+\hat{n}\rangle|-\hat{n}\rangle - |-\hat{n}\rangle|+\hat{n}\rangle\right) = -\frac{1}{\sqrt{2}}e^{i\varphi}\left(|+\hat{z}\rangle|-\hat{z}\rangle - |-\hat{z}\rangle|+\hat{z}\rangle\right)\,.$$

Solution 8.3. Before the transformation, the scalar product of the states is $|\langle k_x|k_y\rangle|^2 = 0$. We can verify that the scalar product after the transformation is $\frac{1}{2}(i-i)=0$, thus the transformation preserves the scalar product and is unitary.

If i is removed from the right-hand side, notice that the transformed states are the same for both $|k_x\rangle$ and $|k_y\rangle$. Thus the scalar product after the transformation would be 1 instead of 0, and the transformation would not be unitary.

Chapter 9

Link with More Traditional Presentations of Quantum Physics

Having reached the end of the book, we feel the need for an additional chapter that would relate our approach to more widespread, traditional approaches to quantum physics. This linking will benefit those who have studied following the traditional method and may be disoriented by the absence of notions like "wave-functions" or "operators". It will benefit the beginners as well, who should at least have an idea of how much there is left to learn, if they wish to go beyond this introduction and reach a serious level of competence in the field.

Contrary to the previous ones, this chapter is *not self-contained*: many notations and notions will be given without a proper explanation. Take it rather as a guide to orient yourself in the material of one of the many excellent textbooks that follow a more traditional approach.

9.1 The Content of Traditional Approaches

In a nutshell, traditional approaches differ from ours in two main respects:

(I) *Choice of the degree of freedom.* Our approach is centered on a very simple example of a physical degree of freedom, namely polarization. As we have seen in the previous chapter, the same mathematics apply to any "two-level system", i.e. a degree of freedom such that only two orthogonal states can be defined. There exist of course systems with more than two orthogonal states, but as long as there are finitely many of these, the main tools are similar. However, extremely important degrees of freedom like *position and momentum* define systems with *infinitely many* orthogonal states. In these cases, while the deep mathematical structure is still retained (in particular, the set of states is still a vector space), the manipulation requires some specific tools. These degrees of freedom, with the corresponding mathematical tools, are normally taught first in traditional approaches. The notions of the *wave-function* and *Heisenberg's uncertainty relations* appear here.

(II) *Importance of dynamics.* In our text we have focused on the preparation and the measurement of quantum systems. Only a very limited number of physical phenomena can be described in this framework: essentially the ones we presented. In order to unravel the full potential of quantum physics, one has to discuss *dynamics*, i.e. the evolution in time of physical systems due to their mutual interaction. It's in the study of dynamics that objects like *Planck's constant* and *Schrödinger's equation* appear, that one can start speaking of energy levels, transition probabilities and similar notions.

We devote the next two sections to each of these topics.

9.2 Description of Systems with Position and Momentum

We have seen that any measurement of polarization yields only two possible results, which we denoted as $+$ and $-$, hence we were led to define two orthogonal states $|+\rangle$ and $|-\rangle$. We could try to follow the same procedure for a measurement of *position* (for simplicity of notation, we consider here a one-dimensional problem). A measurement of position can in principle yield any $x \in \mathbb{R}$ as a result: we should therefore associate a state $|x\rangle$ to each of these measurement outcomes. The gist of the idea is indeed here, but this simple road leads to mathematical problems. The way out, as often in science, consists in moving to a slightly loftier mathematical level.

9.2.1 *Position, wave-functions*

In polarization, we have seen that each state $|\psi\rangle$ is associated to a vector with two components. Therefore, let us declare that "being a vector" is the essential feature: now we just have to find vectors with continuously many components! This sounds very weird at first, indeed one would not know how to write an array with continuously many entries. But vectors are not defined as arrays — you don't define a mathematical object by the way you draw it on a piece of paper. Mathematicians define vectors as elements of a set on which two operations are defined: addition between vectors and multiplication with a number. This abstract definition gives almost immediately the answer to our question: the "vectors" we are looking for are just *functions*! Indeed, if $f(x)$ and $g(x)$ are two functions, the object $h(x) = f(x) + g(x)$ is also a function; if α is a number, the object $\alpha f(x)$ is also a function. Thus functions are vectors in the mathematician's language.

We know now that the state $|\psi\rangle$ of a quantum system possessing position as a degree of freedom can be written as a function $\psi(x)$. It turns out that these functions must be complex, i.e. $\psi(x)$ is a complex number. The scalar product of these vectors is defined as

$$\langle\phi|\psi\rangle = \int_{-\infty}^{\infty} \phi^*(x)\psi(x)\,dx\,. \tag{9.1}$$

Note that this is not unlike the usual scalar product of vectors, if you think of x as the index of the component: we are just multiplying vectors component-wise and taking the sum of these numbers. Now, along with the vector structure, we want to keep the probabilistic interpretation: $|\langle\phi|\psi\rangle|^2$ must still be the probability of finding $|\phi\rangle$ given that the state is $|\psi\rangle$. But integrals over arbitrary functions can go to infinity, while probabilities must be smaller than 1, so this requirement puts a restriction on the functions that one can use to describe quantum states. It turns out that it is necessary and sufficient that these functions satisfy $\int_{-\infty}^{\infty} |\psi(x)|^2 dx = 1$; this set of functions still defines a vector space for the operations above.

At this point, we can explain why the simple road sketched at the beginning was leading to trouble. The state "particle at position x_0" should be described by a function that is zero everywhere but at x_0. The so-called delta-function $\delta(x - x_0)$ would be the only candidate, but it fails to meet the requirements since $\int_{-\infty}^{\infty} \delta(x - x_0)^2 = \infty$. So, strictly speaking, there is no quantum state associated to an infinitely precise location. This mathematical fact is unrelated to the uncertainty relations described on the next page; also, it does not lead to any deep problem in the theory.

A remark on terminology. The functions that represent quantum states are called *wave-functions* for essentially historical reasons: their dynamics are governed by an equation that is analog to those that govern the evolution of waves (see page 131). Only some particular interpretations of quantum physics try to find a physical meaning in the "wave": most physicists are rather convinced that the analogy is only formal. The name has stuck, though, and is very much in use.

9.2.2 *Adding momentum; uncertainty relations*

For the moment, we have described a system that has the degree of freedom "position". We could have done the same construction for "momentum" and we would have gotten functions $\tilde{\psi}(p)$. Now, in order to describe both degrees of freedom, one would be tempted to write $\psi(x,p)$. This is one of the most forbidden mistakes in quantum physics! In this text, we cannot give sufficient reasons to explain why, so we limit ourselves to state things as they are.

Basically, position and momentum cannot be treated separately: only together do they define the degrees of freedom of a physical system. Therefore, a system with position and momentum is described by a single state $|\psi\rangle$. The choice of using the representation $\psi(x)$ or the representation $\tilde{\psi}(p)$ is similar to the choice of a basis to represent a vector; but $\psi(x)$ and $\tilde{\psi}(p)$ describe the *same* state, i.e. contain the same information on the properties of the system. Just as, for ordinary vectors in space, two bases are related by a rotation, these two representations are related by a transformation called the *Fourier transform.*

This mathematical situation translates into the physical fact that position and momentum are incompatible quantities. This fact is extremely well-known and often misquoted. Since the number of incorrect statements is virtually infinite, it is pointless to discuss them here; let us just state the correct version: there is no state of a particle such that both position and momentum are sharply defined. More quantitatively, let Δx be the spread of the wave-function in x and Δp be the spread of the wave-function in p, then

$$\Delta x \, \Delta p \geq \frac{\hbar}{2}, \qquad (9.2)$$

where $\hbar = h/2\pi$ and h is Planck's constant.

9.3 Dynamics of Quantum Systems

In our text, we have said very little about dynamics. In section 1.4, we mentioned that dynamics are assumed to be reversible for isolated systems, just as the case in classical physics; chapter 7 was dealing with a case of a non-isolated system, but the time dependence was admittedly put there by hand without any justification. Traditional approaches, on the contrary, bring dynamics in the play from the very beginning.

9.3.1 *Schrödinger equation: description*

As usual, dynamics are described by a differential equation. The evolution equation of quantum physics is the celebrated *Schrödinger equation*

$$i\hbar\frac{d}{dt}|\psi(t)\rangle = H|\psi(t)\rangle\,. \tag{9.3}$$

The crucial object in the equation is the *Hamilton operator* H, sometimes called the "Hamiltonian". Before turning our attention to it, let us highlight that the equation is a *first-order* differential equation in time. In very elementary texts, this is sometimes stressed as being fundamentally different from Newton's equation, which is second-order in time. But there is no difference, and in fact the evolution equation of the state must be first order in time. If this statement seems strong, just remember the definition of the "state": the mathematical object that describes all the physical properties of the system at a given time. In particular, the state at time $t = 0$ must contain all the meaningful initial conditions. Indeed, the classical state of a particle at time t is given by $S(t) = [x(t), p(t)]$, and it is a standard exercise to show that Newton's equation translates as a first-order equation for $S(t)$.

Now, if H was an ordinary number, the solution of such an equation would be well-known:

$$|\psi(t)\rangle = e^{-iHt/\hbar}|\psi(0)\rangle \tag{9.4}$$

where we used $1/i = -i$. But H is not a number. For finite-dimensional systems, i.e. when $|\psi\rangle$ is a vector of d components, H would be a $d \times d$ *matrix*. For the case of position and momentum, it is a generalized version of a matrix, called an *operator*.

The Hamilton operator is not typical of quantum physics: quantum dynamics was built upon the formalism of classical mechanics as built by Lagrange and Hamilton (precisely) during the nineteenth century. For a particle with position and momentum subjected to a conservative force, the Hamilton operator is just the sum of the kinetic and potential energies:

$$H = \frac{p^2}{2m} + V(x)\,. \tag{9.5}$$

The same form is kept in quantum physics; however, the objects p and x are no longer numbers, but operators (generalized matrices) themselves. If one chooses to work in the x representation of states $\psi(x)$, for instance, x itself becomes a number, but p becomes $-i\hbar\frac{d}{dx}$ (the derivative is obviously an operator, i.e. an operation that can act on a function), thus $p^2 = -\hbar^2 \frac{d^2}{dx^2}$.

An interesting remark: for many systems, the Hamilton operator has the same form as the energy, as in the preceding example; therefore sometimes it is loosely said that the Hamilton operator *is* the energy. But even when the two objects coincide formally, their meaning is very different: the energy is a property of the system, while the Hamilton operator governs the evolution. Moreover, there are cases where the Hamilton operator and energy do not coincide.

9.3.2 *Schrödinger equation: solution*

There is no reason to explain in detail here how an equation involving an operator can be solved: this is exactly what traditional approaches to quantum physics focus on, so we can safely refer to those. As we did before, let us just state the facts without any proof. We focus on Hamilton operators that do not depend on time; if H depends on time, as is the case in some meaningful problems (e.g. an atom driven by a laser), what we discuss here does not apply directly.

The main ideas are simpler to state keeping in mind the finite-dimensional case, where H is a matrix. In order to solve the differential equation, one takes a step that has apparently nothing to do with it: *diagonalize* H, that is, find its eigenvalues and the corresponding eigenvectors. Because of some properties that H must satisfy, the eigenvalues are real numbers and the eigenvectors form a basis.

Why does this help? Suppose that $|\phi_k\rangle$ is an eigenvector of H for the eigenvalue E_k: formally, this means that $H|\phi_k\rangle = E_k|\phi_k\rangle$. Suppose that the initial state of the system is precisely $|\psi(0)\rangle = |\phi_k\rangle$. Then, we can replace the operator H with the *number* E_k in the Schrödinger equation and we immediately have the solution as in Equation (9.4):

$$|\psi(t)\rangle = e^{-iE_k t/\hbar}|\phi_k\rangle\,. \tag{9.6}$$

But we said that the eigenvectors form a basis, therefore any initial condition can be written as

$$|\psi(0)\rangle = \sum_{k=1}^{d} c_k(0)|\phi_k\rangle \tag{9.7}$$

for some components $c_k(0) = \langle\phi_k|\psi(0)\rangle$. By the linearity of the Schrödinger equation, we immediately have the evolution of the most general state:

$$|\psi(t)\rangle = \sum_{k=1}^{d} c_k(t)|\phi_k\rangle \text{ with } c_k(t) = c_k(0)e^{-iE_k t/\hbar}. \tag{9.8}$$

This procedure generalizes to the infinite-dimensional case.

Now, this may sound suspiciously simple: we are claiming that we have the general solution of the Schrödinger equation, while in all textbooks it is stressed that this equation is very difficult, if not sometimes impossible, to solve. But there is no contradiction between the two statements: we have *assumed* that we know the eigenvalues and eigenvectors of H, i.e. that we have solved

$$H|\phi\rangle = E|\phi\rangle. \tag{9.9}$$

This is the hard part, for which indeed there is no general recipe in the infinite-dimensional case. If you manage to solve this linear algebra problem, the solution of the differential equation comes indeed for free according to the scheme sketched here. This is the reason why "diagonalizing Hamiltonians" is the most typical (and feared) exercise in quantum theory.

The following solvable cases are to be found in all textbooks:

- Free particle, i.e. $H = \frac{p^2}{2m}$.
- Square well (the most trivial model of an atom, exhibiting discrete energy levels), square step, square barrier (used to present the famous *tunneling effect*).
- Harmonic oscillator, i.e. $H = \frac{p^2}{2m} + \frac{1}{2}m\omega^2 x^2$.
- Hydrogen atom, i.e. an electron subjected to the Coulomb potential of a proton.

Unfortunately, in each of these cases, Equation (9.9) is solved using a different method. Already, the next simple atom, Helium, is unsolvable in an analytical way: there are now two electrons subjected not only to the Coulomb attraction of the nucleus, but also to their mutual Coulomb repulsion, and this additional term is sufficient to spoil everything.

9.4 Summary

A system that has the degrees of freedom "position" and "momentum" can be described by $\psi(x)$ and $\tilde{\psi}(p)$ respectively. However, both contain the same information on the properties of the system and cannot be sharply defined as described in the uncertainty principle. The Schrödinger equation describes the evolution of quantum states in time, and the solution of the Schrödinger equation involves diagonalizing the *Hamilton operator* H.

9.5 The Broader View

9.5.1 *The original Einstein-Podolski-Rosen argument*

Readers who have some knowledge of the quantum description of position and momentum, and in particular of the corresponding operators, can appreciate an explanation of the original EPR paper [A. Einstein, B. Podolski, N. Rosen, Phys. Rev. **47**, 777 (1935)].

As is well-known, the position q and the momentum p of a particle are incompatible physical quantities. EPR noticed that, by taking two particles, an interesting arrangement can be found. Let $q_A = q \otimes I$ and $p_A = p \otimes I$ be the position and momentum of the first particle, where I is the identity operator, and similarly, let $q_B = I \otimes q$ and $p_B = I \otimes p$ be the position and momentum of the second particle. Using $[q_k, p_k] = i\hbar I$, it is easy to check that the quantities $(q_A - q_B)$ and $(p_A + p_B)$ are compatible. In particular, one can construct a state that is an eigenstate of both: for instance,

$$(q_A - q_B)\psi = x_0\psi \text{ and } (p_A + p_B)\psi = 0\,. \tag{9.10}$$

Let us stress what these conditions mean. There are two particles, each of which can be found everywhere in space, but if the first particle is found at position x_A, the second particle will necessarily be found at position $x_B = x_A - x_0$. Similarly, the momentum of each particle can take any value, but if the momentum of the first is found to be p, the momentum of the second will necessarily be $-p$. Note that the uncertainty relations are not violated by this reasoning: if the measurements are repeated, both x_A and p will be different and actually, since they could take any value, one would have $\Delta q_k = \Delta p_k = \infty$.

Now it is pretty obvious to derive the EPR reasoning. Suppose the position of the first particle is measured: then we also learn the position of the second one (or more precisely, the position where the second one would have been found). But on the second particle, we can measure its momentum, and by doing so we learn also the momentum of the first particle. This way, we know the position and momentum of both particles!

This is a compelling reasoning! What is wrong? The answer is subtle.

One can say that nothing is wrong, in the following sense: those results can indeed be produced with local variables! Indeed, they could come from the lists $\lambda_A = \{x, -p\}$, $\lambda_B = \{x + x_0, p\}$. This should not bother us too much: we know from our study of Bell's inequality (chapter 4.6) that the results of *some* measurements on entangled states can be reproduced with pre-established agreement. And once we look at things from this angle, we also know what is wrong in the general conclusion drawn by EPR: there are other measurements on the same state, or other states for the same measurements, such that the local variable explanation will not hold. Now, if one proves once that local variables do not exist, then they do not exist, even if some specific measurements on some specific states may be reproduced with such models.

It's unfortunate that EPR missed the point. Even more unfortunately, Bohr's reply in defense of the standard interpretation of quantum physics was a showcase of dogmatism that did not address the real issue. Some people speculate that, if Bohr had started a more constructive discussion, Bell's inequality could have been discovered by him or some of his contemporaries, maybe by Einstein himself. What we know for sure is that things did not happen that way, and when Bell discovered the inequality, both Einstein and Bohr were no longer present to continue the debate.

9.6 References and Further Reading

Suggested textbooks that follow a more traditional approach:

- L.E. Ballentine, *Quantum Mechanics: A Modern Development* (World Scientific, 1998).
- J.J. Sakurai, *Modern Quantum Mechanics* (Addison Wesley, 1994).

Concluding Remarks

Congratulations for having reached the final pages of this book. We hope that this book has succeeded in giving you an introduction to basic concepts in quantum physics. Here, we summarize some of the key points that we hope the reader will remember.

- First, quantum systems can exhibit *truly random behavior*. Not everything in the world can be predicted or explained as classical physics may lead you to believe. Playing with Einstein's famous words, it seems that God *does* play dice.
- Second, *entanglement* can occur in composite quantum systems: although the individual states are not well-defined, the global states are. Entangled systems can exhibit surprising correlations that defy all classical explanation.
- Because of the above, we know by now that *nature itself is weird* and behaves non-intuitively at the level of quantum states. This is the reason why quantum physics may be conceptually difficult to comprehend, and also the reason why it should be known beyond the academic circle: its weirdness is not just something that will be explained away by the next smart theorist, it is a fundamental feature of our universe.
- From this book, you may also have noted that these strange concepts of quantum physics, by the very virtue of their "strangeness", have found *potential applications* in interesting fields such as quantum cryptography and random number generation. Some versions of science-fiction phenomena such as cloning and teleportation have found their reality in quantum systems. Quantum physics has not only afforded us better insights into how nature really behaves, it has also opened up new possibilities in what technology can achieve.

There is much more to quantum physics than what we have presented in this book, many more surprises, interesting applications and discoveries. We hope that your journey into the quantum world does not end here, that this book is just your first step into this field.

Index